A MicroCase® Workbook for

SOCIAL RESEARCH

Michael Corbett
BALL STATE UNIVERSITY

Lynne Roberts

THIRD EDITION

WADSWORTH
™
THOMSON LEARNING

Australia • Canada • Mexico • Singapore • Spain • United Kingdom • United States

Sociology Editor: *Lin Marshall*
Assistant Editor: *Analie Barnett*
Editorial Assistant: *Reilly O'Neal*
Technology Project Manager: *Dee Dee Zobian*

Marketing Manager: *Matthew Wright*
Production Service: *Jodi Gleason*
Copy Editor: *Margaret Moore*
Printer: *Transcontinental Printing, Inc.*

For more information, contact
Wadsworth/Thomson Learning
10 Davis Drive
Belmont, CA 94002-3098
USA

For more information about our products, contact us:
Thomson Learning Academic Resource Center
1-800-423-0563
http://www.wadsworth.com

International Headquarters
Thomson Learning
International Division
290 Harbor Drive, 2nd Floor
Stamford, CT 06902-7477
USA

UK/Europe/Middle East/South Africa
Thomson Learning
Berkshire House
168-173 High Holborn
London WC1V 7AA
United Kingdom

Asia
Thomson Learning
60 Albert Complex, #15-01
Singapore 189969

Canada
Nelson Thomson Learning
1120 Birchmount Road
Toronto, Ontario M1K 5G4
Canada

ISBN 0-534-58189-7

Contents

About the Authors

Michael Corbett received his Ph.D. from the University of Iowa. He is now Professor of Political Science at Ball State University, where he teaches courses covering an introduction to political science, research methods in political science, and public opinion. He is author of numerous scholarly articles and of four books: *Political Tolerance in America: Freedom and Equality in Public Attitudes* (New York: Longman, 1982); *American Public Opinion: Trends, Processes, and Patterns* (New York: Longman, 1991); *Research Methods in Political Science: An Introduction Using MicroCase* (Belmont, CA: Wadsworth Group/Thomson Learning, 2001); and *Politics and Religion in the United States* (New York, NY: Garland Press, 1999), which was coauthored with his wife, Julia.

Lynne Roberts received her Ph.D. from Stanford University where she also was on the staff of the Computing Center. She left Stanford to inaugurate a year-long research methods course in the then-new doctoral program offered by the School of Social Welfare at the University of California, Berkeley. After several years, she joined the faculty of the University of Washington in Seattle where she served as Associate Professor of Sociology. In 1984, Roberts left academic life to pursue her interests in educational software. After leaving academia, she served as president of MicroCase Corporation. During her tenure at Washington, Roberts taught statistics at both the undergraduate and graduate levels. Trained as an experimental social psychologist, she designed and conducted many experiments, the results of which she reported in a book and a number of scholarly articles. She also has published articles on methodological subjects including scaling and experimental design.

Getting Started

INTRODUCTION

This workbook is about *doing* social research. There is nothing make-believe about what you will be doing. You'll use the same resources and techniques used by professional social science researchers.

Since the exercises differ in length and difficulty, your instructor may assign only selected questions, or you may be given alternative questions. Be sure to carefully check any information from your instructor so that you will do the appropriate problems. Some of the problems will ask you to print certain results and attach them to your worksheet. If you do not have a printer or you have been instructed not to use a printer, simply complete the written exercises.

Many of the exercises will have a variety of possible answers. This reflects the nature of research—there is no "correct" way to do research. In most situations, a variety of approaches are equally appropriate.

The exercises are designed so that the introductory material does not require a computer. If your computer time is limited, you may read this material beforehand and save your computer time for completing the worksheets. In any case, you should carefully read the introductory material before starting on the written exercises.

SYSTEM REQUIREMENTS FOR STUDENT MICROCASE

- Windows 95 (or higher)
- 8 MB RAM
- CD-ROM drive
- 3.5" disk drive
- 15 MB of hard drive space (if you want to install it on the hard drive)

You can run Student MicroCase in three different ways:

- Run it directly from the CD-ROM and diskette without installing it.
- Install it on a hard drive and run it from there.
- Run it from a network version of Student MicroCase.

NETWORK VERSIONS OF STUDENT MICROCASE

A network version of Student MicroCase is available at no charge to instructors who adopt this book for their course. Note that Student MicroCase can be run directly from the CD and diskette on virtually any computer network—regardless of whether a network version of Student MicroCase has been installed.

INSTALLING STUDENT MICROCASE

Note: If you will be running Student MicroCase directly from the CD-ROM or if you will be using a version of Student MicroCase that is already installed on a network, then skip to the section "Starting Student MicroCase."

To install Student MicroCase, you need both the diskette and the CD-ROM that are packaged inside the back cover of this book. Then follow these steps in order:

1. Start your computer and wait until the Windows desktop is showing on your computer.
2. Insert the floppy diskette into the A drive (or B drive) of your computer.
3. Insert the CD-ROM disc into the CD-ROM drive.
4. On most computers, the CD-ROM will automatically start in a few seconds and a welcome menu will appear. If the CD-ROM doesn't automatically start, do the following:

   Click on [Start] from the Windows desktop, click [Run], type **D:\SETUP**, and click [OK].

   Note: If the CD-ROM drive is not the D drive, substitute the actual letter of the CD-ROM drive for the D in D:\SETUP above. For example, if the CD-ROM drive is the E drive, the command would be **E:\SETUP**.

5. To install Student MicroCase to your hard drive, select the second option on the list: "Install Student MicroCase to your hard drive."
6. During the installation, you will be presented with several screens (described below). In most cases, you will be required to make a selection or entry and then click [Next] to continue.

The first screen that appears is the **License Name** screen. (If this software has been previously installed or used, it already contains the licensing information. In such a case, a screen confirming your name will appear instead.) Here you are asked to type your name. It is important to type your name correctly, since it cannot be changed after this point. Your name will appear on all printouts, so make sure that you spell it completely and correctly! Then click [Next] to continue.

A **Welcome** screen then appears. This provides some introductory information and suggests that you shut down any other programs that might be running. Click [Next] to continue.

The **Software License Agreement** appears next. Read this screen and click [Yes] if you accept the terms of the software license.

The next screen has you **Choose the Destination** for the program files. You are strongly advised to use the destination directory that is shown on the screen. Then click [Next] to continue.

The Student MicroCase program will now be installed. At the end of the installation, you will be asked if you would like a shortcut icon placed on the Windows desktop. It is recommended that you select [Yes]. You are then informed that the installation of Student MicroCase is finished.

Click the [Finish] button and you will be returned to the opening Welcome screen. To exit completely, click the option "Exit Welcome Screen."

INSTALLING STUDENT MICROCASE TO A LAPTOP COMPUTER

If you are installing Student MicroCase to a hard drive on a laptop that has both a CD-ROM drive and a floppy disk drive, simply follow the preceding instructions. However, if you are installing Student MicroCase to a hard drive on a laptop where you cannot have both the CD-ROM drive and floppy disk drive attached at the same time, follow these steps in order:

1. Attach the CD-ROM drive to your computer and insert the CD-ROM disc.

2. Start your computer and wait until the Windows desktop is showing on your screen.

3. On most computers the CD-ROM will automatically start and a welcome menu will appear. If it does, click [Exit].

4. Click [Start] from the Windows desktop, select [Programs], and select [Windows Explorer].

5. Click the drive letter for your CD-ROM in the left column (usually D:\). A list of folders and files on the CD-ROM will appear in the left column.

6. From the Windows Explorer menu, click [Edit] and [Select All]. The folders and files on the CD-ROM will be highlighted. Using your mouse, right click (use your right mouse button) on the list of folders and files. From the box that appears select [Copy].

7. In the left column of the Windows Explorer menu, right click once on your C drive (do NOT select a folder) and select [Paste] from the box that appears.

8. Close Windows Explorer by clicking the [X] button on the top right corner.

9. Remove the CD-ROM and CD-ROM drive from your computer and attach the floppy disk drive. Place the floppy disk from your workbook in the drive.

10. Click [Start] from the Windows desktop, click [Run], type C:\SETUP, and click [OK].

11. Select the first option from the Welcome menu: **Run Student MicroCase from the CD-ROM**. Within a few seconds Student MicroCase will appear on your screen.

STARTING STUDENT MICROCASE

As indicated previously, there are three ways to run Student MicroCase: (1) directly from the CD-ROM and diskette; (2) from a hard drive installation; or (3) from a network installation. Each method is described below.

STARTING STUDENT MICROCASE FROM THE CD-ROM AND DISKETTE

Unlike most Windows programs, you can run Student MicroCase directly from the CD-ROM and diskette without installing any part of it. To do so, follow these steps:

1. Insert the 3.5" diskette into the A or B drive of your computer.

2. Insert the CD-ROM disc into the CD-ROM drive.

3. On most computers the CD-ROM will automatically start and a welcome menu will appear in a few seconds.

   Note: If the CD-ROM does not automatically start after it is inserted, click [Start] from the Windows desktop, click [Run], type D:\SETUP, and click [OK]. If your CD-ROM drive is not the D drive, replace the letter D with the proper drive letter.

4. Select the first option from the Welcome menu: **Run Student MicroCase from the CD-ROM**. Within a few seconds, Student MicroCase will appear on your screen.

STARTING STUDENT MICROCASE FROM A HARD DRIVE INSTALLATION

If Student MicroCase is installed to the hard drive of your computer (see earlier section "Installing Student MicroCase"), it is not necessary to insert either the CD-ROM or floppy diskette. Instead, locate the Student MicroCase "shortcut" icon on the Windows desktop, which looks something like this:

To start Student MicroCase, position your mouse pointer over the shortcut icon and double-click (that is, click it twice in rapid succession). If you did not choose to place the shortcut icon on the desktop during installation of Student MicroCase (or if the icon has been accidentally deleted), you can start Student MicroCase as follows:

- Click [Start] from the Windows desktop.

- Click [Programs].

- Click [MicroCase].

- Click [Student MicroCase].

After a few seconds, Student MicroCase will appear on your screen.

STARTING STUDENT MICROCASE FROM A NETWORK

If the network version of Student MicroCase has been installed to a computer network, you must insert the 3.5" floppy diskette (not the CD-ROM) that comes with your book. Then double-click the Student MicroCase icon that appears on the Windows desktop to start the program. (Note: Your instructor may provide additional information that is unique to your computer network.)

If Student MicroCase is installed on a network, but the icon has been accidentally deleted from the Windows desktop of the computer you are using, you can start Student MicroCase as follows:

- Click [Start] from the Windows desktop.

- Click [Programs].

- Click [MicroCase].

- Click [Student MicroCase].

After a few seconds, Student MicroCase will appear on your screen.

STUDENT MICROCASE MENUS

Student MicroCase is extremely easy to use. All you do is "point and click" your way through the program. That is, use your mouse arrow to point at the selection you want, then click the left button on the mouse.

Two menus provide the beginning points for everything you will do in Student MicroCase. When you start Student MicroCase, the **FILE & DATA MENU** appears first. In order to do statistical analysis, you switch to the **STATISTICS MENU**. You can toggle back and forth between these two menus by clicking the menu names shown on the left side of the screen.

Not all options on a menu are always available. You cannot, for example, do statistical analysis until you have a data file open. You can tell which options are available at any given time by looking at the colors of the options. The options that are not available at a particular time are dimmed. For example, when you first start Student MicroCase, only the OPEN FILE and the NEW FILE options are immediately available. As you can see, the colors for these options are brighter than those for the other options shown on the screen. Further, when you move your mouse pointer over the options that are available at a particular time, they become highlighted.

SOFTWARE GUIDES

Throughout this workbook, there are "Software Guides" that provide you with the basic information needed to carry out each task. Here is an example:

> ➤ *Data File:* **USA**
> ➤ *Task:* **Mapping**
> ➤ *Variable 1:* **101) MURDER**
> ➤ *View:* **Map**

Each line of this Software Guide is actually an instruction. Let's follow the simple steps to carry out this task.

STEP 1: SELECT A DATA FILE

Before you can do almost anything in Student MicroCase, you need to open a data file.

To open a data file, click the OPEN FILE task from the **FILE & DATA MENU**. A list of data files will appear in a window (e.g., GSS, USA, etc.). If you click on a file *once*, a description of the highlighted file is shown in the window next to this list. In the Software Guide shown above, the ➤ symbol to the left of the *Data File* step indicates that you should open the USA data file. To do so, click on [USA] and then click the [Open] button (or just double-click on [USA]). The next window that appears (labeled *File Settings*) provides additional information about the data file, including a file description, the number of cases in the file, and the number of variables, among other things. To continue, click the [OK] button. You are now returned to the main menu of Student MicroCase. (You won't need to repeat this step until you want to open a different data file.) Notice that you can always see which data file is currently open by looking at the file name shown on the top line of the screen.

STEP 2: SELECT A TASK

Once you have selected a data file, the next step is to select a program task. First, switch to the **STATISTICS MENU**. There are eight analysis tasks offered in this version of Student MicroCase. Not all tasks are available for each data file because some tasks are appropriate only for certain kinds of data. MAPPING, for example, is a task that applies only to ecological data, and thus cannot be used with survey data files.

In the Software Guide we're following, the ➤ symbol on the second line indicates that the MAPPING task should be selected, so click the MAPPING option with your left mouse button.

STEP 3: SELECT A VARIABLE

After a task is selected, you will be shown a list of the variables in the open data file. Notice that the first variable is highlighted and a description of that variable is shown in the Variable Description window at the lower right. You can move this highlight through the list of variables by using the up and down cursor keys (as well as the <Page Up> and <Page Down> keys). You can also click once on a variable name to move the highlight and update the variable description. Go ahead—move the highlight to a few other variables and read their descriptions.

If the variable you want to select is not showing in the variable window, use the scroll bars located on the right side of the variable list window to move through the list. See the following figure:

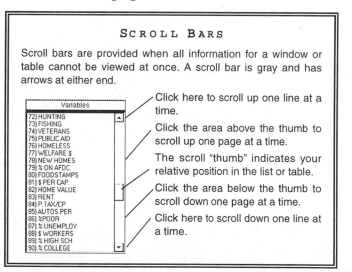

By the way, Appendix A at the back of this workbook contains a list of the variable names for key data files provided in this package.

Each task requires you to select one or more variables, and the Software Guides indicate which variables should be selected. The Software Guide example here indicates that you should select 101) MURDER as Variable 1. On the screen, there is a box labeled Variable 1. Inside this box there is a vertical cursor that indicates that this box is currently an active option. When you select a variable, it will be placed in this box. Before selecting a variable, be sure that the cursor is in the appropriate box. If it is not, place the cursor inside the appropriate box by clicking the box with your mouse. This is important because in some tasks the Software Guide will require more than one variable to be selected, and you want to be sure that you put each selected variable in the right place.

To select a variable, use any one of the methods shown below. (Note: If the name of a previously selected variable is in the box, use the <Delete> or <Backspace> key to remove it—or click the [Clear All] button.)

- Type in the **number** of the variable and press <Enter>.

- Type in the **name** of the variable and press <Enter>. Or you can type just enough of the name to distinguish it from other variables in the data—MUR would be sufficient for this example.

- Double-click on the desired variable in the variable list window. This selection will then appear in the variable selection box. (If the name of a previously selected variable is in the box, the newly selected variable will replace it.)

- Highlight the desired variable in the variable list, then click the arrow that appears to the left of the variable selection box. The variable you selected will now appear in the box. (If the name of a previously selected variable is in the box, the newly selected variable will replace it.)

Once you have selected your variable (or variables), click the [OK] button to continue to the final results screen.

STEP 4: SELECT A VIEW

The next screen that appears shows the final results of your analysis. In most cases, the screen that first appears matches the "view" indicated in the Software Guide. In this example, you are instructed to look at the Map view—that's what is currently showing on the screen. In some instances, however, you may need to make an additional selection to produce the desired screen.

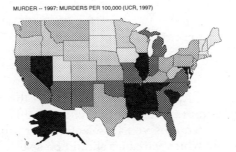

MURDER -- 1997: MURDERS PER 100,000 (UCR, 1997)

(OPTIONAL) Step 5: Select an additional Display

Some Software Guides will indicate that you should select an additional "Display." In that case, simply click on the option indicated for that additional display. For example, this Software Guide might have included an additional line that required you to select the Legend display.

Step 6: Continuing to the Next Software Guide

Some instructions in the Software Guide may be the same for at least two examples in a row. For instance, after you display the map for murder in the example above, the following Software Guide might be given:

> Data File: **USA**
>> Task: **Mapping**
> ➤ Variable 1: **122) CRIME RATE**
>> ➤ View: **Map**

Notice that the first two lines in the Software Guide do not have the ➤ symbol located in front of the items. That's because you already have the USA data file open and you have already selected the MAPPING task. With the results of your first analysis showing on the screen, there is no need to return to the main menu to complete this next analysis. Instead, all you need to do is select CRIME RATE as your new variable. Click the [🖾] button located in the top left corner of your screen. The variable selection screen for the MAPPING task appears again. Replace the variable with 122) CRIME RATE and click [OK].

To repeat: You only have to do those items in the Software Guide that have the ➤ symbol in front of them. If you start from the top of the Software Guide, you're simply wasting your time.

If the Software Guide instructs you to select an entirely new task or data file, then you will need to return to the appropriate menu. If you have been using one task (e.g., MAPPING), simply click the [Menu] button located at the top left corner of the screen to return to the **STATISTICS MENU** to select a new task. If you need to open a different data file, switch to the **FILE & DATA MENU**.

That's all there is to the basic operation of Student MicroCase. Just follow the instructions given in the Software Guide and point and click your way through the program.

On-Line Help

Student MicroCase has extensive on-line help. You can obtain task-specific help by pressing <F1> at any point in the program. For example, if you are using correlation, you can press <F1> to see the help for the **CORRELATION** task.

If you prefer to browse through a list of the available help topics, select **Help** from the pull down menu at the top of the screen and select the **MicroCase Help**

Topics option. At this point, you will see a list of topic areas. Each topic is represented by a closed-book icon. To see what information is available in a given topic area, double-click on a book to "open" it. (For this version of the software, use only the "Student MicroCase" section of help.) When you double-click on a book graphic, a list of help topics is shown. A help topic is represented by a graphic with a piece of paper with a question mark on it. Double-click on a help topic to view it.

If you have questions about Student MicroCase, try the on-line help described above. If you are not very familiar with software or computers, you may want to ask a classmate or your instructor for assistance.

EXITING FROM STUDENT MICROCASE

If you are continuing to the next section of this workbook, it is *not* necessary to exit from Student MicroCase quite yet. But when you are finished using the program, it is very important that you properly exit the software—do not just walk away from the computer or remove your diskette. To exit Student MicroCase, return to the main menu and select the [Exit] button that appears on the screen.

Important: If you inserted your CD-ROM disc and/or floppy diskette prior to starting Student MicroCase, remember to remove them before leaving the computer.

Introductory Exercise: Exploring Data Files

Overview

In this brief, introductory exercise, you will discover how easy it is to use MicroCase to explore two of the data files that you will be using throughout this workbook. Make sure you have already gone through the *Getting Started* section that is provided in the previous section of the workbook. It contains important information that is essential for doing this and future exercises.

In this exercise you will learn to

- use the USA data file to map variables;
- use the GSS data file to create pie charts and bar graphs;
- use the search feature to find any variables that contain a particular word or phrase;
- print the results of your analysis.

There is no chapter in the textbook that corresponds to this introductory exercise. Thus, you may complete this exercise before you begin reading the textbook.

MicroCase Data Files: A Quick Tour

In the *Getting Started* section, you learned how to follow the "Software Guides" in order to complete the required analysis task. Let's take this a step further and look at several additional features of Student MicroCase and explore some of the data files that you will be using. Other features in Student MicroCase will be explained later as you need them.

MAPPING

Follow the instructions in the MicroCase Guide below.

➤ *Data File:* **USA**
➤ *Task:* **Mapping**
➤ *Variable 1:* **129) HUNTING**
➤ *View:* **Map**

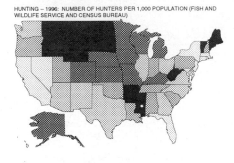

HUNTING -- 1996: NUMBER OF HUNTERS PER 1,000 POPULATION (FISH AND WILDLIFE SERVICE AND CENSUS BUREAU)

> **Below the Software Guide, you will sometimes be provided additional information or tips about how to create the requested analysis. This information will appear in bold, like it is here. For this example, remember that the first line indicates that you should open the USA data file, the second line tells you to select the MAPPING task, the third line indicates that you should select 129) HUNTING as Variable 1, and the final line indicates that your final view should be the map.**

Note that the full description of the 129) HUNTING variable is as follows: NUMBER OF HUNTERS PER 1,000 POPULATION. A variable is anything that varies among the objects being examined. Since we are dealing with states in the USA data file, a variable here would be something that varies across states. For example, the crime rate is one such variable. Church membership rate, voter participation rate, population, and geographic area are other variables included in this data file. In this course, we'll spend quite a bit of time talking about variables, how to create them, and how to use them. Here we will see how this variable, 129) HUNTING, varies across the states.

At this point you have a map of the United States on the screen similar to the one shown above. The map shows the hunting rate for each state. The states appear in five different colors, from very light to very dark. The darker the state, the higher the hunting rate. As you can see, there are patterns here. In general, the hunting rates are higher in the northwest and the upper midwest.

Let's explore some of the other options on the screen by using the following Software Guide.

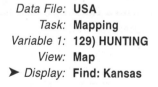

Data File: **USA**
Task: **Mapping**
Variable 1: **129) HUNTING**
View: **Map**
➤ Display: **Find: Kansas**

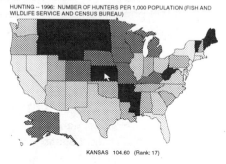

HUNTING -- 1996: NUMBER OF HUNTERS PER 1,000 POPULATION (FISH AND WILDLIFE SERVICE AND CENSUS BUREAU)

KANSAS 104.60 (Rank: 17)

Notice that only the last step has a ➤ symbol located in front of it. That means that the first three steps are not required if you are continuing from the previous example. With the hunting map showing on the screen, you need only to click the [Find case] option. A list of the states will appear. Select Kansas from the list. Then click [OK]. The map for Kansas has now been highlighted and information about it is presented on the screen.

We see that Kansas has 104.60 hunters per 1,000 population and is ranked 17th in the nation in terms of this variable.

You can also see the information for a particular state simply by clicking on that state on the map. Try this with several states. Locate some states that are ranked very high on this hunting rate by clicking on states that have very dark colors. Also locate some states that are ranked very low by clicking on states that have very light colors.

To see all 50 states ranked from highest to lowest, we need to use MicroCase's [List: Rank] option.

Data File: **USA**
Task: **Mapping**
Variable 1: **129) HUNTING**
➤ View: **List: Rank**

RANK	CASE NAME	VALUE
1	Wyoming	282.74
2	South Dakota	252.03
3	Montana	221.59
4	West Virginia	203.75
5	Idaho	201.79
6	Vermont	179.36
7	Mississippi	157.34
8	Maine	156.75
9	Arkansas	149.33
10	North Dakota	137.93

Again, notice that only the final step must be completed if you are continuing from the previous example. In this case, click the [List: Rank] option.

You can see from this that Wyoming is highest with a rate of 282.74 hunters per 1,000 population and New Jersey is lowest with a rate of 11.71. Click on the [Map] option to return to the map and then do the final step in this Software Guide.

Data File:	**USA**
Task:	**Mapping**
Variable 1:	**129) HUNTING**
➤ *View:*	**Map**
➤ *Display:*	**Legend**

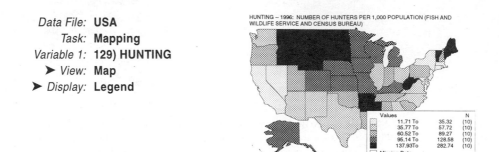

The legend shows you what the colors on the map represent. You can see, for example, that states with hunting rates between 137.93 and 282.74 are in the highest group, while states with rates from 11.71 to 35.32 are in the lowest group. To remove the legend from the screen, simply click the [Legend] option again.

Student MicroCase offers a second way to view the map, which you can see by completing the final step shown in this Software Guide.

Data File:	**USA**
Task:	**Mapping**
Variable 1:	**129) HUNTING**
View:	**Map**
➤ *Display:*	**Spot**

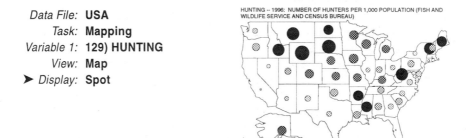

Each state's hunting rate is represented by a spot. The size of the spot is determined by the relative magnitude of the value of each state for the hunting rate variable. Thus, states that have high hunting rates, such as Wyoming and South Dakota, have large, dark-colored spots. State that have low hunting license rates, such as Florida and New Jersey, have small, light-colored spots.

Let's look at the map of another variable. Follow the instructions in the Software Guide below, but as you do, remember that you can select a variable in any one of several ways: by typing in the name of the variable, by typing the number of the variable, or by selecting the variable from the variable list. Try this using each of the three methods.

Data File: **USA**
Task: **Mapping**
➤ Variable 1: **94) PICKUPS**
➤ View: **Map**

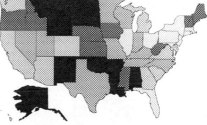

PICKUPS -- 1995: PICKUP TRUCKS PER 1,000 POPULATION (HIGHWAY)

If you do not remember how to return to the variable selection screen, you should probably review the *Getting Started* section again. Remember, it is not necessary to return to
the main menu since you do not need to select a new data file or analysis task.

This is a map of the number of pickups per 1,000 population. Let's rank the
states to see which states have the highest rate of pickup trucks.

Data File: **USA**
Task: **Mapping**
Variable 1: **94) PICKUPS**
➤ View: **List: Rank**

RANK	CASE NAME	VALUE
1	Wyoming	231
2	Montana	208
3	Alabama	189
4	Arkansas	187
5	North Dakota	185
6	New Mexico	182
6	South Dakota	182
8	Oklahoma	178
9	Alaska	173
10	Louisiana	163

We see that Wyoming ranks the highest with a rate of 231 pickups per 1,000
population. This isn't much of a surprise. In fact, the map of pickup trucks looks
very similar to the hunting rate map.

Student MicroCase allows you to print all results that you obtain. There is a
printer icon on the tool bar. If you click the print button, a window will appear
that gives you the option to print out the result on the screen. (You can also
modify the printer settings, if necessary). Fancy graphics, such as maps, take
longer to print than, say, a table or a page of text. Throughout this workbook, you
will be required to print out results and turn them in with your worksheet assign-
ments. Incidentally, when you print out a result from Student MicroCase, your
name and the current date will appear at the top of each page. This allows you to
easily locate your printout if you are using a shared printer.

Let's look at another map.

Data File: **USA**
Task: **Mapping**
➤ *Variable 1:* **79) WARM WINTR**
  ➤ *View:* **Map**

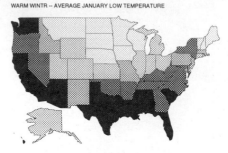

WARM WINTR -- AVERAGE JANUARY LOW TEMPERATURE

This is a map of the average low temperature in January. This map is almost exactly the opposite of the map of hunting rates. Not too surprising—the best opportunities for hunting tend to be in the colder states. If you select the [List: Rank] option, you will see that Hawaii is the warmest state with an average January low of 65, while North Dakota is the coldest with an average of –3 degrees.

Now let's see how easy it is to compare two maps in MicroCase.

Data File: **USA**
Task: **Mapping**
Variable 1: **79) WARM WINTR**
➤ *Variable 2:* **129) HUNTING**
  ➤ *Views:* **Map**
  ➤ *Display:* **Spot**

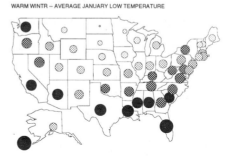

WARM WINTR -- AVERAGE JANUARY LOW TEMPERATURE

$r = -0.509$**

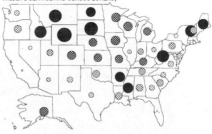

HUNTING -- 1996: NUMBER OF HUNTERS PER 1,000 POPULATION (FISH AND WILDLIFE SERVICE AND CENSUS BUREAU)

Note that you do not have to modify Variable 1 (WARM WINTR) if you are continuing from the previous example. But you do need to select HUNTING as Variable 2. To do so, click once in the Variable 2 box to make it active, then select the HUNTING variable as you normally would. When the map appears, make sure to select the spot display as indicated.

You can now compare the map of winter temperatures with the map of the hunting rate.

Let's look at one last map before we leave this data file.

Data File: **USA**
Task: **Mapping**
➤ *Variable 1:* **20) KID ABUSES**
➤ *View:* **Map**

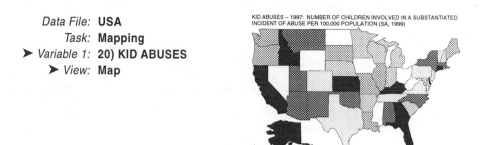

KID ABUSES -- 1997: NUMBER OF CHILDREN INVOLVED IN A SUBSTANTIATED INCIDENT OF ABUSE PER 100,000 POPULATION (SA, 1999)

This map shows the substantiated child abuse rate (the number of children involved in a substantiated incident of child abuse per 100,000 population) by state. Notice that on this map there are seven states (e.g., Wyoming and Ohio) that are not assigned one of the colors. Let's list the values for each state so that we can see what is going on with these seven states.

Data File: **USA**
Task: **Mapping**
Variable 1: **20) KID ABUSES**
➤ *View:* **List: Rank**

RANK	CASE NAME	VALUE
1	Alaska	1468.57
2	Kansas	707.19
3	Idaho	673.96
4	Georgia	595.45
5	Delaware	593.55
6	Connecticut	555.22
7	Florida	534.90
8	California	533.17
9	Kentucky	528.02
10	Arizona	514.14

If you scroll down to the bottom of the list, you will find the seven states that have no regular color on the map also have no value listed for this variable. No data were available for this variable for these seven states. Hence, they have what is called "missing data." It is not unusual to have missing data for a variable, and this sometimes causes problems for statistical analysis.

UNIVARIATE STATISTICS

Next we will explore a different data file—one based on responses from individuals in a survey—and we will look at some different MicroCase features. As shown in the Software Guide below, open the GSS data file and select the UNIVARIATE task.

➤ *Data File:* **GSS**
 ➤ *Task:* **Univariate**

These data are from the 1998 General Social Survey conducted by the National Opinion Research Center. There were 2,832 respondents (or cases), and this data file contains over 100 variables from that survey. Let's look at some of the variables in this data file. The first variable concerns how often the respondent reads the newspaper. When the categories of a variable are labeled, these labels are shown in the variable description window. Thus, for example, the second variable (WATCH TV), hours spent watching television per day, has four categories: 1 or less, two, 3–4, and over 4. The next variable (EVER UNEMP), concerning whether the respondent has ever been unemployed, has just two categories: No and Yes.

By the way, you probably noticed that the variable descriptions for the GSS data file have what appears to be another variable name in parentheses at the end of the description. For example, the variable description for variable 2) WATCH TV reads: "On the average day, about how many hours do you personally watch television? (TVHOURS)." What does TVHOURS mean here? This is the original name given to the variable by the National Opinion Research Center (NORC). We have used somewhat different names here (especially when we altered the variable in some way), but we have included the original names for those who might want to look up the variable in NORC materials. Otherwise, you can simply ignore these variable names in parentheses.

Scroll through the list of variables and highlight any that look of interest to you. Examine each variable description to obtain further information about the variable. Some of these variables relate to characteristics of the respondents such as age, sex, and race. These are called *demographic* characteristics. Other variables are about attitudes and behaviors, such as attitudes toward abortion and voting behavior.

Also note that the variables in this data file are based on individual people rather than on geographic areas such as states. Thus, we cannot use the MAPPING task for individuals because you cannot "map people." So we will use the UNIVARIATE task to begin our look at these data. Let's select the variable SEX, as shown in this Software Guide.

Data File: **GSS**
Task: **Univariate**
➤ *Primary Variable:* **89) SEX**
➤ *View:* **Pie**

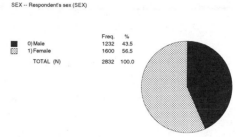

A pie chart appears and a legend for the pie colors is shown at the left. We can see that the slice of the pie representing females is larger than the slice representing males. (In Chapter 4 of the text, we will see that the sample used in this survey has a slight sex bias: females are somewhat overrepresented in the sample relative to their percentage of the general population.)

Note that this screen with the pie chart also presents other information about the variable. In the legend on the left side of the screen, you can see that the category *Male* has a 0 next to it and the category *Female* has a 1 next to it. Computers are more efficient working with numbers than with words, so researchers usually assign a number code to each category. For the moment, just ignore these number codes and focus on the category labels (male and female).

The frequency results show that 1,232 of the respondents are male and 1,600 are female. In terms of percentages, 43.5 percent are male and 56.5 percent are female. Note that the sum of these percentages will be 100 percent, although the sum will sometimes deviate slightly from 100 percent simply because of rounding errors. Now let's look at these results in a bar graph.

Data File: **GSS**
Task: **Univariate**
Primary Variable: **89) SEX**
➤ *View:* **Bar - Freq.**

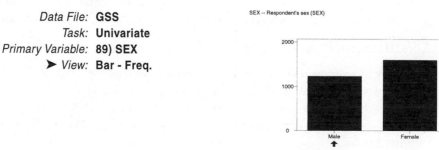

As indicated by the ➤ symbol, if you are continuing from the previous example, select the [Bar - Freq.] option.

These results are shown in a bar graph, where each bar represents a category. The information shown below the bar graph represents the one bar that has an arrow directly beneath it. To see the values for the other bar, simply click on that bar with your mouse. Here again, we see that the bar for females is somewhat bigger than the bar for males, indicating that there are more females than males in this sample.

Let's look at another variable in the GSS data file.

> Data File: **GSS**
> Task: **Univariate**
➤ Primary Variable: **59) WOMAN PRES**
➤ View: **Pie**

In this survey, 1,687 respondents said that they would be willing to vote for a woman for president, 116 respondents said that they would not, and there are 1,029 missing cases. In computing the percentages, the missing cases are excluded. Thus, 93.6 percent *of those who answered the question* said that they would be willing to vote for a woman for president while only 6.4 percent *of those who answered the question* said they would not be willing.

Missing data can be a problem in social research, but there actually is not a problem here with the 1,029 missing cases. In this survey, some questions were asked of only part of the sample (randomly selected). Only about two-thirds (1,871 respondents) of the sample were asked this particular question. Thus, of those who were asked this particular question, only 68 people didn't answer it.

SEARCHING FOR VARIABLES IN A DATA FILE

Return to the variable list and let's explore the Search feature. If we wanted to find a variable dealing with the religious preference of the respondent, it might take some time to scroll through the list of variables and find this particular variable. However, we can search for such a variable quite easily. This applies to any of the MicroCase tasks in which there is a list of variables.

When you have the list of variables on the screen, click on the [Search] option below the list. The box that appears will instruct you to type in a word, phrase, or character for which to search. You type in a word or phrase that might appear in either the variable name or the variable description or a variable for which you are looking.

In this example, you want to search for a variable about the religious preferences of people. Thus, you could use the term *religious preference* for a search term. Type in **religious preference** and click [OK]. As you can see, MicroCase found three variables dealing with religious preference. The first one, 74) RELIGION, is the one you want as the other two deal with the strength of religious preference.

In this particular situation, the use of the term *religious preference* narrowed the search very well. However, there are situations in which you don't want to restrict the search too much. If you wanted to find other variables in this data file about religion, you might want to search for the term *relig*. MicroCase would find any variable in the data file that contained the string *relig* anywhere in the variable name or the variable description. This procedure, for example, would select any variable that contained the word *relig*ious or the word *relig*ion.

Let's do this search. First, click on the [Full List] button to clear the results from the previous search and return to the full list of variables. Then, click the [Search] button and edit the old search term so that *relig* appears as the new search term. Click [OK] to see the results of your new search. You now see a list of twelve variables that relate to religion in some way. You can view the variable descriptions from this abbreviated list of variables and select any of these variables for analysis, as you did above with the full list of variables.

After you have finished with the [Search] option, return to the full list of variables by clicking on [Full List].

The worksheet section that follows will give you a further introduction to Student MicroCase.

NAME: _____

COURSE: _____

DATE: _____

You will need to use Student MicroCase to complete the worksheets. It is essential that you have first gone through the *Getting Started* section and the preliminary part of this exercise.

1. Let's begin by examining another data file. The NES file contains selected variables from the 1998 American National Election Study. There were 1,281 respondents in this sample. Open the NES data file and select the UNIVARIATE task as shown below.

> ➤ *Data File:* **NES**
> ➤ *Task:* **Univariate**

Scroll to variable 105) NO PARTIES and examine the variable description. Write down the variable description and category labels below.

a. 105) NO PARTIES
Description:

_____

_____

List the category labels:

    1. _____

    2. _____

    3. _____

b. Now do the same thing for the variable 45) KNOW HOUSE.
Description:

_____

_____

List the category labels:

0. _____

1. _____

2. Use the [Search] feature to find variables in the NES data file that use each of the phrases listed below. (For example, in the first situation below, you would search for any variables containing the word *abortion*.)

   a. Abortion
      List the number and name of each variable found in the search.

   b. Representative
      List the number and name of each variable found in the search.

   c. Foreign
      List the number and name of each variable found in the search.

3. a. So far, you have been looking at the variable names and descriptions. Now you will obtain univariate statistics for two variables. First, carry out the following task and fill in the category labels, the frequency in each category (the number of cases in each category), and the percentage of cases in each category.

Data File:	**NES**
Task:	**Univariate**
➤ *Primary Variable:*	**66) PTY CONTRL**
➤ *View:*	**Pie**

CATEGORY LABELS	FREQUENCY	PERCENT
_____	_____	_____
_____	_____	_____
_____	_____	_____

For this variable, how many cases were missing data? _____

b. Now obtain the univariate statistics for the variable 2) AGE CATEGR, and fill in the category labels, frequencies, and percentages below.

CATEGORY LABELS	FREQUENCY	PERCENT
_____	_____	_____
_____	_____	_____
_____	_____	_____
_____	_____	_____

For this variable, how many cases were missing data? _____

4.   Now open the USA data file so you can explore it further.

> ➤ *Data File:* **USA**
> ➤ *Task:* **Mapping**
> ➤ *Variable 1:* **38) TEEN MOMS**
> ➤ *View:* **Map**

a. Write the variable description for 38) TEEN MOMS below.

_____

_____

b. Look at the map for this variable and click on the darkest-colored states. Which region of the country has the highest rate of teenage motherhood? (Circle one.)

Northeast

Midwest

South

West

c. As indicated by this abbreviated Software Guide, use the [List: Rank] option to see the rankings of states on this variable

▶ *View:* **List: Rank**

	NAME	RATE
Highest State	_____	_____
Next Highest State	_____	_____
Third Highest State	_____	_____

d. What are the rates for the following states? (Hint: Select the option for listing cases alphabetically.)

	RATE
North Carolina	_____
Indiana	_____
California	_____

e. Print out the ranked distribution for the TEEN MOMS variable—if you switched to the alphabetical list in the previous question, make sure to switch back to the ranked order before printing. (Note: If your computer is not connected to a printer or if you have been instructed not to use the printer, skip these printing instructions.)

5. For each of the following variables, map the variable, use the [List: Rank] option, and then specify the three highest and the three lowest ranked states (omit states with missing values).

a. 121) VIOLENT CR

Three Highest States	Three Lowest States
_____	_____
_____	_____
_____	_____

b. 11) SUICIDE

Three Highest States	Three Lowest States
_____	_____
_____	_____
_____	_____

6. a. Let's compare two maps.

> Data File: **USA**
> Task: **Mapping**
> ➤ Variable 1:   **121) VIOLENT CR**
> ➤ Variable 2:   **6) %URBAN**
> ➤ Views: **Map**

Compare the two maps visually and then look at the rankings. Are these two maps similar to one another (darker states on one map are darker on the other, lighter states on one map are lighter on the other), opposite of one another (darker states on one map are lighter on the other, and vice versa), or neither? (Circle one.)

Similar   Opposite   Neither

b. Now compare the map of 121) VIOLENT CR with the maps for each of the variables listed below, and indicate in each situation whether the two maps are similar, opposite, or neither. Circle the correct answer.

107) AUTO THEFT   Similar   Opposite   Neither

18) %COLLEGE   Similar   Opposite   Neither

73) %VOTED 96   Similar   Opposite   Neither

7. a. Let's look at the map for the percentage of the population age 25 and over that has a college degree

> Data File: **USA**
> Task: **Mapping**
> ➤ Variable 1:   **18) %COLLEGE**
> ➤ View: **Map**

b. The percentage of the population with a college degree might be related to the income level of states. Search for variables concerning income, select one of these variables, and write the variable number and variable name below.

Variable number and name: _____

c. Compare the map showing the percentage of the population that has a college degree with the map of the income variable you selected. Are the two maps similar to one another, opposite, or neither? (Circle one.)

Similar

Opposite

Neither

That's all for this exercise. Remember to exit Student MicroCase properly (return to the main menu and click the [Exit] button). If you inserted your floppy diskette or CD-ROM disc at the beginning of the session, remember to remove it once you have exited Student MicroCase.

1

Concepts and Theories

OVERVIEW

In this exercise, you will learn more about how we use concepts and theories in social research. The exercise also emphasizes the ability to recognize tautologies and the ability to identify testable statements. In addition you will learn to recognize normative statements based on values.

BEFORE YOU BEGIN

Please make sure that you have read Chapter 1 in the textbook and can answer the following review questions (you need not write any answers):

1. How are concepts and theories related to one another?

2. What is the difference between description and explanation?

3. What principles guide the construction of good concepts in social research?

4. What is a tautology?

5. What are the two essential features of theories in science? *ABSTRACT, CORRELATION*

6. Can theories be proven? Why or why not?

7. In relation to concepts, what are indicators? *observed measure of a concept.*

8. How is the test of a hypothesis used to assess a theory?

9. What is the difference between induction and deduction?

10. What are empirical generalizations and how do they relate to induction?

11. What does it mean to say that there is a correlation between two variables?

ABSTRACT - How and why concepts are linked
Correlation - positive/negative relationship between indicators, vary in unison

Theoretical concepts do not stand alone but, rather, are embedded in a set of relationships with other concepts. Concepts with the same name may have quite different meanings in different theories. Let's look at an example.

Sociologists have long been interested in social class, yet this concept means different things to different people. The Marxist perspective defines *social class* as the relationship of the individual to the means of production. Within this framework, there are two classes: those who own the means of production and those who don't. Other theoretical approaches divide societies into a greater number of classes, each of which can be distinguished clearly from the others.

Still other sociologists prefer to use the term *social status* since they see social status as a continuum rather than a set of identifiable classes. Even those who see social class as a continuum disagree on how it should be operationalized. Some see social class as a property of the individual, while others define it as a property of the individual's occupation. If social status, or class, is a property of the individual, then two individuals with the same occupation might have different social status. If social status is a function of the individual's occupation, then individuals with different incomes and educational levels but the same occupation will have the same social status.

None of these is the *true* definition of social class, and each definition suggests different empirical definitions. These definitions (and theories) will, however, differ in how useful they are in explaining the empirical world.

A concept should not be confused with a theory. For example, William Sims Bainbridge[1] observed an interesting phenomenon in his field study of Satan's Power, a satanic cult. When the cult first formed, members had friends outside the group as well as inside the group. As time passed, their social contacts outside the group began to decrease. This happened not only because some of the friends became members of the group, but also because members tended to stop seeing friends who were outside the cult. Eventually, members of the group had no contacts outside the group at all. Bainbridge called this phenomenon *social implosion* to describe how social relationships collapsed in upon themselves. However, naming this phenomenon added nothing to our understanding. We still don't know if this event was unique to this group or, if not, under what conditions members of groups tend to break off outside contacts or even what types of groups are likely to experience social implosion. Interesting as this phenomenon may be, we simply have a concept in need of a theory.

Confusing a concept with a theory is a relatively common problem. The difference between the two is that a theory is testable, or falsifiable, while a concept is not. We cannot say that social implosion caused this group to cut off outside contacts. Such a statement would be a tautology: Social implosion causes social implosion. On the other hand, the statement "The greater the punishment for nonconformity to group norms, the greater the likelihood that members of the group will cease contacts with outsiders" is, in principle, testable. "The more

---

[1] Bainbridge, William Sims. 1978. *Satan's Power*. Berkeley: University of California Press.

deviant the norms of the group from the norms of society, the more likely that members of the group will cease contacts with outsiders" is similarly, in principle, testable. The "in principle" caution means that we may encounter practical problems if we attempt to test the idea. For example, we may not be able to find enough appropriate groups, we may not be able to adequately measure the concept, we may encounter ethical problems, and so on.

Let's consider a second, more complicated example. Robert K. Merton[2] created the following table:

		Uses Socially Approved Means	
		Yes	No
Seeks Socially Approved Goals	Yes	Conformist	Innovator
	No	Ritualist	Retreatist

This may appear to be a theory because two different concepts are involved, socially approved means and socially approved goals. However, all we have are four different definitions. A *conformist*, for example, is defined as one who seeks socially approved goals and uses socially approved means—anyone else is not a conformist. An *innovator* is defined as one who uses means that are not socially approved to seek socially approved goals—if this is not true of the individual, then the individual is not an innovator. Each of the two remaining types are defined in a similar manner. This is called a *typology* because a combination of two characteristics is used to define a set of types.

Assuming we have information on the means used and the goals desired by a particular individual, we can place him or her in the appropriate category or type. The statement "John Dillinger was an innovator *because* he used means that were not socially approved (robbing banks) to seek goals that were socially approved (money)" is a *tautology*—it is merely the application of the definition of innovator to John Dillinger. It is not a testable statement.

If this typology were linked to some other concept, then it would be more than a set of definitions. For example, the statement "Innovators will be more common under democracies than under other forms of government" and the statement "Innovators are more likely to obtain material rewards than are conformists" are both, in principle, testable.

Now, let's take a look at a theory and how it might be tested. Most theories of crime have focused on the criminal and the characteristics that lead individuals into a life of crime. Lawrence Cohen and Marcus Felson developed an approach to deviance called *opportunity theory*. They realized that having individuals with a propensity to commit crime may be necessary but is not sufficient for a crime to occur. There also needs to be an opportunity—without banks, there can be no bank robbers. A crime requires not only a person motivated to commit the offense,

---

[2] Merton, Robert K. 1938. "Social Structure & Anomie." *American Sociological Review*, 3:672–82.

but also the presence of a suitable target (property or individual victims) and the absence of effective guardians. For example, this theory would predict that burglaries of homes will be more likely to occur during the daytime when everyone is at school or work (the absence of effective guardians) than in the evening when homes are occupied (the presence of effective guardians).

This is a testable statement. We can check the empirical truth or falsity of this statement. If it is false, we should reject the theory.[3] If it holds true, then we can have greater faith in the theory. We can then conduct additional tests of this theory. Some empirical hypotheses provide stronger tests of theories than do others. In general, the riskier the prediction, the stronger the test. Suppose, for example, that the empirical prediction is well known to be true in advance of any research. Since we expect a theory to be consistent with known facts, this would not be a very strong test of the theory. However, if the empirical hypothesis is the opposite of what most social scientists would expect, then this would be a much stronger test of the theory.

Often a researcher may find support for a particular hypothesis and read into this additional, scientifically unjustified, conclusions. Consider a researcher who confirms the hypothesis that children who are spanked model the behavior of their parents and consequently engage in interpersonal violence. Beyond reporting this finding, the researcher adds that this study therefore proves that children should not be spanked. This is a *normative* conclusion, a statement that prescribes how people should behave based on some underlying set of values. That is, the conclusion does not logically follow from the assumptions but requires an additional value statement: Interpersonal violence is bad. Groups who value aggressive behavior might reach exactly the opposite conclusion. Normative statements cannot be evaluated in terms of true or false but only in terms of right or wrong and thus are not testable.

---

[3] In some cases, rather than reject the theory, we might reject the empirical evidence as an inadequate test of the theory. In this example, if a neighborhood has hired a security force to patrol during the daytime, then the theory would not predict a higher burglary rate during the day.

NAME: _____

COURSE: _____

DATE: _____

Workbook exercises and software are copyrighted. Copying is prohibited by law.

EXERCISE

1

WORKSHEET

1. For each of the following statements, circle **Yes** if the statement is a tautology or **No** if the statement is not a tautology.

    a. When it's your time to die, you die.   (Yes)   No

    b. People with higher education are more politically tolerant
       because they are exposed to greater social diversity.   Yes   (No)

    c. Robert is self-centered because his parents always gave him
       everything he wanted and let him do whatever he wanted.   Yes   (No)

    d. Cocaine possession is illegal because there are laws against it.   (Yes)   No

    e. Mary votes Republican because she has a very high income.   Yes   (No)

    f. Hernando won the election because he received a majority of
       the votes cast.   (Yes)   No

    g. Ellie has a hereditary disease because she was born that way.   (Yes)   No

    h. Most people become more cautious in terms of lifestyle as they
       grow older.   Yes   (No)

    i. Girls are more cooperative than boys because they are treated
       differently in their families and by others in society.   Yes   (No)

2. For each of the following statements, circle **T** if the statement is, in principle, testable. Circle **NT** if the statement is not testable. Then, since the testability of a statement may depend on how you interpret certain elements, explain your answer in one sentence.

    a. Higher income people are less supportive of governmental
       programs to assist the poor than lower income people are.   (T)   NT

    b. The use of marijuana should be made legal.   T   (NT)

23

c. Women are more likely to be Democrats than men are.   (T)   NT

d. Political activists take part in political activities.   T̸   (NT)

e. Democracy is the best form of government.   T   (NT)

f. The crime rate is higher in areas where unemployment is high. (T)   NT

3. In social research, we often use education as a variable, and we usually
   operationalize education as the number of years of schooling an individual
   completes. If we conceptualize education as the amount of learning
   achieved by an individual, there are other ways that we might operationalize this concept (although these ways would not be very practical in most
   social research).

   a. If we conceptualize education as the amount of learning achieved by an
      individual, what are some problems with using years of schooling to
      serve as an indicator of education?

   PART TIME College EDUCATION = 8 yrs ⎫
   Full Time College ED = 4 yrs ⎭ same degree Achieved

b. Describe an alternate method of operationalizing education that would better reflect the amount of learning achieved.

Assoc. Degree, Bach, Masters etc...

or

level 1, 2, 3, 4

4. The concept of *civil religion* refers to the use of religious symbols and rhetoric for public purposes and functions, especially political functions. *Civil religion* prompts U.S. presidents to mention God in the inaugural speeches. This also is the reason those in charge of designing U.S. money place "In God We Trust" on coins.

Based on materials in the textbook and the earlier discussion in this exercise, evaluate the last two sentences in the above paragraph. Does civil religion explain why U.S. presidents mention God and why "In God We Trust" is placed on coins? Discuss your answer.

yes- in order to appeal to the masses/christian america the word "God" is added.

5. a. Give two additional empirical predictions from the opportunity theory of crime developed by Cohen and Felson.

1. *An unlocked car with valuables left visible is more likely to be broken into than a locked car with no visible valuables*

2. *A person who is dressed like he is wealthy, and exhibits the actions of a wealthy person, is more likely to be robbed than one who is less dressed less extravagantly*

b. Assuming you have great confidence in this theory, how could you apply the ideas to crime prevention programs?

Advise citizens to lock cars, hide valuables and take care not to draw to much attention to themselves when in public

6. Consider the concept *personal trust*. Please give a nominal definition of this concept and then provide an operational definition for it.

Nominal Definition: (What does personal trust *mean*?)

The ability to have faith in a persons reliability, responsibility and moral standards. The value of a persons moral being

Operational Definition: (How do you *measure* personal trust? Include at least two specific questions that could be used to measure personal trust.)

Can a person hold down a responsible job?

~~Have~~ Has a person ever lost his driving privileges?

Criminal record?

7. These remaining questions require you to use Student MicroCase. (It is assumed that by this point you have completed the "Getting Started" section and the "Introductory Exercise.") Start Student MicroCase, open the GSS data file, and look at the variable descriptions for variables 21) WORK IMPOR and 57) WOMEN HOME. Note that these questions are based on normative statements and that we cannot prove or disprove these statements scientifically. However, in social research we often study the views that people hold on such matters.

a. Find one other variable in the GSS file that is based on a normative statement and list it below.

Number:_____   Name:_____

Description:

_____

_____

_____

b. For contrast with the normatively based questions, find two questions that are *not* based on normative statements and list them below.

Number:_____   Name:_____

Description:

_____

_____

Number:_____   Name:_____

Description:

_____

_____

8. For each of the concepts listed below, search through the variables in the GSS file and find one variable to measure the concept. Write the variable number and name where indicated.

CONCEPT	**VARIABLE NUMBER AND NAME**
Religious preference	_____
Political party preference	_____
Confidence in the military	_____
Gun ownership	_____

2a

The Research Process Using Aggregate Data

OVERVIEW

In this exercise, you will learn more about the stages of the social research process and you will experience some of these stages directly. Specifically, you will gain experience in formulating hypotheses, selecting indicators to operationalize concepts, and using statistical results to test hypotheses. Two exercises have been designed to parallel Chapter 2 in the textbook: Exercise 2a and Exercise 2b. As the above title suggests, this first exercise focuses on the use of aggregate data. The second exercise deals with survey data. You will see the differences in the ways we utilize the two types of data as you make your way through these exercises.

BEFORE YOU BEGIN

Please make sure you have read Chapter 2 in the textbook and can answer the following review questions (you need not write any answers):

1. What are the general stages in the research process?

2. What does *operationalize a concept* mean, and how are *indicators* used in this process?

3. What is replication research and why is it important?

Many of us are concerned about violent crimes, and we wonder why some people are more likely than others to commit such crimes. Learning theorists would argue that violent criminal activities are learned in the same way that other behaviors are learned. According to this approach, to understand violent behavior, we need to see how the behavior is learned. This has led some social scientists to speculate that there should be a connection between hunting and violent crime: Hunting encourages the learning of violent behaviors, and, once learned, these behaviors can be applied to humans as well as to animals.

At this point, we have completed two steps in the research process: **select our topic** (Step 1) and **formulate our research question** (Step 2). The research question is simply "Is there a connection between hunting and violent behavior?" Next we need to **define the concepts** (Step 3):

Hunting: the pursuit and killing of wild, game animals for sport or for food

Violent behavior: physical violence, or threat of physical violence, toward another individual

Both of these concepts apply only to behaviors outside of a regular occupational role. Killing animals in an occupational role—what meat packers, chicken farmers, and humane society workers do—is not considered hunting. Further, for present purposes, let's exclude the threat of violence by a police officer or in a military setting from our definition of violent behavior.

When we define concepts, we also introduce the possibility that others will disagree with the definition. Because we are interested in hunting behavior, not in its motivation, we have included both hunting for food and hunting for sport in our definition. If we were testing other ideas concerned with motivation for hunting, our definition might exclude hunting for food.

We now need to **operationalize our concepts** (Step 4)—to select the indicators of these concepts. In this exercise, we'll use the **USA** data file to test this idea. The USA data set is based on aggregates, rather than individuals. That is, each case, or state, is a collection, or aggregation, of individuals.

Start Student MicroCase using the instructions in the *Getting Started* section and proceed through the tasks in the following guide—select the USA data file and select the MAPPING task.

> *Data File:* **USA**
>> *Task:* **Mapping**

Based on the first exercise, we already know that we have information on the hunting rate (the number of hunters per 1,000 population) in this data file. This is our indicator of the extent of hunting in each state. Now scroll through the variable descriptions and see if you can find a variable that could be used as an indicator of violent behavior across the states. Locate variable 101) MURDER in your variable list, click on the variable to highlight it, and then read the description in the Variable Description box. This looks like a good indicator of violent behavior.

If the idea that hunting affects violent behavior is true, then we would expect states to be relatively high on both variables or to be relatively low on both variables. In other words, we would expect the maps of the two variables to look alike. We can now **formulate our hypothesis** (Step 5): States with high rates of hunting will tend to have high murder rates. We are now ready to make the observations.

We are fortunate that the appropriate observations have already been collected and are included in the USA data set, so we can skip Step 6, **make the observations**, and move directly to **analyze the data** (Step 7). Let's do that now.

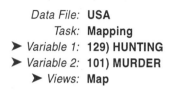

Data File: **USA**
Task: **Mapping**
➤ *Variable 1:* **129) HUNTING**
➤ *Variable 2:* **101) MURDER**
➤ *Views:* **Map**

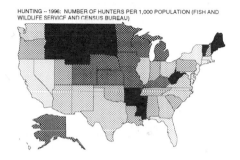

HUNTING -- 1996: NUMBER OF HUNTERS PER 1,000 POPULATION (FISH AND WILDLIFE SERVICE AND CENSUS BUREAU)

r = − 0.252*

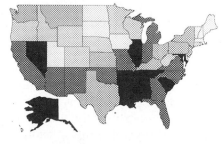

MURDER -- 1997: MURDERS PER 100,000 (UCR, 1997)

Remember that the ➤ symbol indicates which tasks you need to perform. You have already selected the USA data file and the MAPPING task. So here you continue by selecting the first variable. Next you select the second variable, and then you click [OK] to view the comparison of the two maps.

The map of HUNTING is shown at the top of the screen, and the map of MURDER is shown in the lower half of the screen.

The maps don't look at all alike; in fact, they appear to be somewhat opposites. Based on visual inspection, we would conclude that the two rates don't vary together in the predicted direction. We have now finished Step 8—**assess the**

results. (If this were an actual study, we would still have two steps remaining: publish the findings and replicate the research.)

Now let's compare the hunting licenses map with a map showing the circulation of *Field & Stream* magazine.

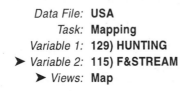

Data File: **USA**
Task: **Mapping**
Variable 1: **129) HUNTING**
➤ Variable 2: **115) F&STREAM**
➤ Views: **Map**

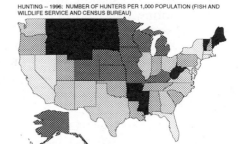

HUNTING -- 1996: NUMBER OF HUNTERS PER 1,000 POPULATION (FISH AND WILDLIFE SERVICE AND CENSUS BUREAU)

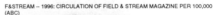

r = 0.923**

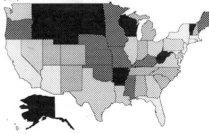

F&STREAM -- 1996: CIRCULATION OF FIELD & STREAM MAGAZINE PER 100,000 (ABC)

These two maps look very much alike. However, conclusions from simply inspecting maps can be fairly subjective; our personal opinions may influence how we view the maps. So it is preferable to be able to quantify how alike or how different two maps are.

Fortunately, there is a statistical technique that will allow us to do this. So let's examine this relationship between hunting licenses and *Field & Stream* circulation differently.

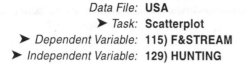

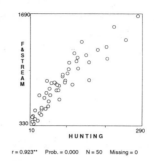

The ➤ on the Task line indicates that you must return to the main menu and select a new task—SCATTERPLOT.

A graphic appears in the middle of the screen. The y-axis is the line at the left labeled F&STREAM—the values on this axis range from 330 to 1690. This vertical axis represents all values of F&STREAM that we saw earlier on the map. The x-axis is the line at the bottom of the graph labeled HUNTING—the values on this axis range from 10 to 290. This horizontal axis represents all values of HUNTING that we also displayed earlier in a map.

Each dot on the graph represents a state and its values on these two variables. Let's see how this works. Wyoming has a value of 1681 on F&STREAM, so it would be very close to the top of the y-axis. In your mind, draw a *horizontal* line at approximately 1681 on the y-axis. Wyoming has a value of 282.74 on HUNTING, so it would be at the far right of the x-axis. Mentally draw a *vertical* line at approximately 283 on the x-axis. The intersection of the two lines represents Wyoming.

Student MicroCase has a feature that allows you to find any case on your scatterplot. Let's use it to locate Wyoming.

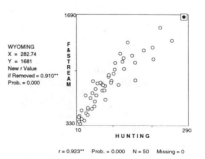

If you are continuing from the previous example, click the [Find] option. An alphabetical list of the states appears in a window. Select Wyoming from the list. Then click [OK].

A highlighted square appears around the dot representing Wyoming on the scatterplot. This dot is located where the two lines would intersect and represents Wyoming on both variables. Further, the screen now presents the value of Wyoming on each variable (listed as X and Y). You might want to find the location of some other states—click on [Find] once to deselect the current case and once more to select a new case. When you have finished, deselect the case and let's look at the regression line.

Data File:	**USA**
Task:	**Scatterplot**
Dependent Variable:	**115) F&STREAM**
Independent Variable:	**129) HUNTING**
➤ *View:*	**Reg. Line**

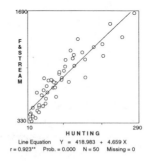

If you are continuing from the previous example, simply click the [Reg. Line] option.

A line now appears on the graph. This line represents the best effort to draw a *straight* line that connects all of the dots. It is unnecessary for you to know how to calculate the location of the regression line—the program does it for you. But if you would like to see how the regression line would look if the maps were identical, all you need to do is examine the scatterplot using the same variable for both the x-axis and the y-axis. All the dots would be on the line.

In the current example, the maps are very similar but not identical, so the dots are scattered near, but not on, the regression line. Now we need a method of determining how close these dots are to the line. With the regression line showing on the scatterplot, select the [Residuals] option.

Data File:	**USA**
Task:	**Scatterplot**
Dependent Variable:	**115) F&STREAM**
Independent Variable:	**129) HUNTING**
➤ *View:*	**Reg. Line/Residuals**

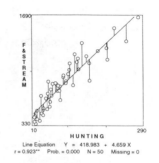

If you are continuing from the previous example, click the [Residuals] option.

A vertical line now connects each dot to the line; if we sum these distances, we can get a measure of how much alike the two maps are. The smaller this sum, the more alike are the two maps, or variables. For example, when the maps are identical and all the dots are on the regression line, the sum of these distances is zero.

This idea is used to calculate the value of a statistic called the *correlation coefficient*. The value of this coefficient can vary from –1 to +1. When the value is 1, the two maps are identical; when the value is –1, they are opposites. When the value is zero, they are neither similar nor opposites. At the bottom of the scatterplot, we can see that the correlation (indicated using a lowercase r) between hunting and circulation of *Field & Stream* magazine is .923. This shows a very strong relationship between these two variables.

When the value of the correlation coefficient gets close to zero, the relationship might simply be due to chance factors, such as inaccurate measurements. How large must a correlation be to indicate that a relationship between the two variables actually exists? Using probability theory, statisticians can tell us how likely we are to observe a particular correlation coefficient by chance when there is really no relationship. Note that there are two asterisks next to the correlation (r = .923**). The two asterisks signify that the probability that this relationship could have occurred just because of chance factors is only 1 chance out of 100 or less. One asterisk would mean that the probability that this relationship could have occurred just because of chance factors is 5 out of 100 or less.

If this probability is small enough, then we can reject the hypothesis of no relationship, thus supporting the hypothesis that there is a relationship. In social science, the level of .05 is used for this rejection point—that is, if this correlation coefficient would be observed less than 5 times in 100 when there is no relationship, then we will conclude that a relationship probably exists. This is called the *statistical significance level*. Some researchers are even more stringent and use .01 as the rejection point. As indicated above, if the correlation reaches this level of significance (or better), then there are two asterisks next to it. You will encounter the significance level again in other types of analyses. But you will see that in some situations you will be given the exact significance level (e.g., Prob. = .013), and other times the level of significance will be indicated only by asterisks. (In Chapter 4 of the text, these issues will be discussed in greater detail.)

Let's go back to our research hypothesis about hunting and murder and look at the scatterplot for those two variables.

Data File: **USA**
Task: **Scatterplot**
➤ Dependent Variable: **101) MURDER**
➤ Independent Variable: **129) HUNTING**
➤ View: **Reg. Line**

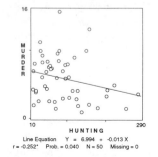

Line Equation   Y = 6.994 + -0.013 X
r = -0.252*   Prob. = 0.040   N = 50   Missing = 0

Remember, it is not necessary to return to the main menu to change your variable selections. Use the 🔄 button to return to the variable selection screen. Then, to replace variables that were previously selected, double-click on the new variables you want to use. Or you can click the [Clear All] button and select new variables at this point.

The regression line slopes *downward* and the correlation (–0.252*) between these two variables is negative. This relationship is significant at the .05 level—there is one asterisk next to the correlation. This means that the empirical data actually go in the opposite direction from that predicted—the higher the hunting rate, the lower the murder rate! At this point, we would want to assess whether we really have provided a test of our research question. Perhaps we did not properly operationalize the concepts. Perhaps the observations are incorrect. There are other such problems that could have influenced the results. Some empirical observations provide much stronger tests of hypotheses than do others. By the end of the course, you should understand why this was not a particularly strong test.

However, for the present, let's assume that this was a strong test of a theory and that the empirical hypothesis was, in fact, a proper application of the theory. What can we conclude about the theory? Finding that the empirical hypothesis was not supported should lead us to reject the theory from which it was derived. The implications of the theory were wrong; therefore the theory itself must be false. On the other hand, if the empirical hypothesis were true, we could not conclude that the theory is *true*. We might have more faith in the theory, but we could not *prove* the theory. (You might want to reread the discussion of the relationship between theory and empirical observations in Chapter 1 of the text to understand this better.)

In the present example, you might find it tempting to decide that we should have used another theoretical approach. For example, you might suggest that hunting is a substitute form of violence and individuals who hunt have no need for other outlets for violent behavior. Creative social scientists could probably come up with many different ideas that would be consistent with this empirical result. However, we can't claim to have tested any of these alternatives, because we constructed them after we already knew the results of the test. Explanations of this type are called *ex post facto*—created after the fact. For research to be a valid test of a theory, the empirical hypothesis must be derived from the theory before the relevant analysis is conducted.

Incidentally, findings that fail to support a hypothesis are also important. In fact, a strong test that raises serious questions about an existing theory may change the whole course of a scientific field. Sometimes taking a step backward is really a leap forward.

The scatterplot technique we have just used is the basis for the first correlation coefficient (developed by Karl Pearson in the 1890s), and there are many variant methods based on the same underlying logic. However, it is not necessary to actually create a scatterplot in order to calculate r (the correlation coefficient), and thus it is possible to calculate many correlations at the same time. Let's use the CORRELATION task to do this.

Data File: **USA**

➤ Task: **Correlation**

➤ Select Variables: **101) MURDER**

129) HUNTING

130) FISHING

94) PICKUPS

Correlation Coefficients
N: 50   Missing: 0
Cronbach's alpha: Not calculated--negative correlations
LISTWISE deletion (1-tailed test)   Significance Levels: ** = .01, * = .05

	MURDER	HUNTING	FISHING	PICKUPS
MURDER	1.000	-0.252*	-0.092	0.007
HUNTING	-0.252*	1.000	0.654**	0.747**
FISHING	-0.092	0.654**	1.000	0.607**
PICKUPS	0.007	0.747**	0.607**	1.000

The variable selection for this task works differently than other tasks you've seen. If you type in a variable name or number in the Select Variables box and press the <Enter> key, the variable name will be placed in the larger box below it. This allows you to select multiple variables for the correlation analysis. To delete a variable from this list, click on the variable name and press the <Delete> key on your keyboard. Of course, you can also use the [Clear All] button to remove all variables that were previously selected.

Looking at the top left cell, we see that there is a perfect correlation (1.000) between MURDER and MURDER, as there should be since these are the same measures. Looking down the diagonal from left to right, we can see that in fact each variable is perfectly correlated with itself. Reading down the far left column, we can see the correlation between the murder rate and each of the other three variables. The correlation with HUNTING is the same as we have seen with the scatterplot. The negative signs indicate that, as each of these rates rises, the murder rate declines. Again, two asterisks indicate that a correlation is statistically significant beyond the .01 level, while one asterisk indicates significance beyond the .05 level. Notice that the correlation between MURDER and PICKUPS (the number of pickup trucks per 1,000 population) lacks an asterisk. That means it is not statistically significant and should be regarded as zero.

In addition to showing the correlation of each variable with MURDER, the matrix shows the correlations between each pair of variables. To find the correlation between any two variables, first find the name of one variable across the top of the table and then find the name of the other down the left side. Locate the cell where the two variables coincide and that is the correlation coefficient between them. Thus, for example, the correlation between HUNTING and PICKUPS is 0.747**, and between HUNTING and FISHING it is 0.654**.

In addition to using the SCATTERPLOT and CORRELATION tasks for obtaining correlations, you can see the correlation between any two aggregate variables by using the MAPPING task. Let's take a look at this.

Data File: **USA**
➤ Task: **Mapping**
➤ Variable 1: **101) MURDER**
➤ Variable 2: **115) F&STREAM**
➤ Views: **Map**

MURDER -- 1997: MURDERS PER 100,000 (UCR, 1997)

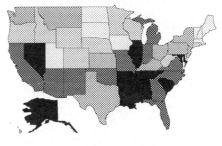

r = – 0.284*

F&STREAM -- 1996: CIRCULATION OF FIELD & STREAM MAGAZINE PER 100,000 (ABC)

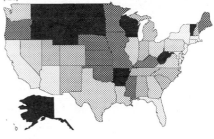

Notice that the two maps tend somewhat to be reverse images, indicating a negative correlation. But instead of relying on visual comparisons of these two maps, look at the value of the correlation coefficient between the two variables (r = –0.284*). Whenever you compare two maps, the correlation automatically appears.

Your turn.

NAME: _____

COURSE: _____

DATE: _____

Workbook exercises and software are copyrighted. Copying is prohibited by law.

EXERCISE

2a

WORKSHEET

1. Crime rates vary from one area to another in the United States, and a variety of factors such as variation in the degree of urbanism might account for this. Here, however, let's examine the link between alcohol consumption and crime. At the individual level, excessive alcohol consumption sometimes contributes to people committing crimes that they would not have committed otherwise. At the aggregate level, we can hypothesize that states that have higher per capita consumption of alcohol will have higher crime rates. Let's test this hypothesis.

> ➤ Data File: **USA**
> ➤ Task: **Mapping**
> ➤ Variable 1:   **122) CRIME RATE**
> ➤ Variable 2:   **52) ALCBEV/CAP**

Compare the two maps. Are they similar to one another, opposite of one another, or neither similar to nor opposite of one another? (Circle one.)

Similar   Opposite   Neither

Write in the value of the correlation coefficient.   r = _0.79_

Is the correlation positive or negative?
(Circle one.)   Positive   Negative

Is the relationship statistically significant? (Circle one.)   Yes   No

Do the results support the hypothesis? (Circle one.)   Yes   No   NO

2. Voting for candidates is an individual matter, and most of the time we want to examine survey data in order to test hypotheses about voting choices. However, we can also discover certain voting patterns by doing aggregate analysis of election data. For example, using the USA data file, we might investigate what kind of states are more likely to vote Democratic in presidential elections and what kinds of states are more likely to vote Republican. Let's examine the map of 70) %CLINTON96.

> Data File: **USA**
> Task: **Mapping**
> ➤ Variable 1:   **70) %CLINTON96**
> ➤ View: **List: Rank**

a. Look at the states that gave Clinton the highest percentages of their votes in the 1996 presidential election. Did Clinton do better in the eastern part of the country or the western part? (Circle one.)

Western part   ~~Eastern part~~

b. The Democratic Party has traditionally been viewed as being more supportive of economically disadvantaged groups. Thus, we could hypothesize that Clinton would do better in states that have more poor people. Let's test this using the SCATTERPLOT task.

> Data File: **USA**
> ➤ Task: **Scatterplot**
> ➤ Dependent Variable: **70) %CLINTON96**
> ➤ Independent Variable: **32) POV LINE**

Write in the value of the correlation coefficient. $r =$ ___1___

Is the correlation positive or negative?
(Circle one.)   ~~Positive~~   Negative

Is the relationship statistically significant? (Circle one.)   Yes   ~~No~~

Do the results support the hypothesis? (Circle one.)   Yes   ~~No~~

c. Perhaps we did not adequately operationalize poverty by using the percentage of the population below the poverty level. So, let's try another indicator to see whether the percentage of the vote for Clinton was higher in states that had more poor people. This time we will use the median household income as an indicator. We hypothesize that there is a negative relationship between the percentage of the vote for Clinton and the median household income in states.

> Data File: **USA**
> Task: **Scatterplot**
> Dependent Variable: **70) %CLINTON96**
> ➤ Independent Variable: **30) MED FAM$**

Write in the value of the correlation coefficient. $r =$ .138

Is the correlation positive or negative?
(Circle one.)   (Positive)   Negative

Is the relationship statistically significant? (Circle one.)   Yes   (No)

Do the results support the hypothesis? (Circle one.)   Yes   (No)

d. Let's hypothesize that Clinton would do better in states that have higher unemployment.

> Data File: **USA**
> Task: **Scatterplot**
> Dependent Variable: **70) %CLINTON96**
> ➤ Independent Variable: **95) UNEMPLOY**

Write in the value of the correlation coefficient. $r =$ ___.238___

Is the correlation positive or negative?
(Circle one.)   (Positive)   Negative

Is the relationship statistically significant? (Circle one.)   (Yes)   No

Do the results support the hypothesis? (Circle one.)   ~~Yes   No~~ yes

e. Let's hypothesize that Clinton would do better in states that have higher rates of their citizens receiving TANF (Temporary Assistance for Needy Families) funds.

> Data File: **USA**
> Task: **Scatterplot**
> Dependent Variable: **70) %CLINTON96**
> ➤ Independent Variable: **37) TANF FAM**

Write in the value of the correlation coefficient. $r =$ ___.552___

Is the correlation positive or negative?
(Circle one.)   (Positive)   Negative

Is the relationship statistically significant? (Circle one.)   (Yes)   No

Do the results support the hypothesis? (Circle one.)   (Yes)   No

3.   We might expect that the percentage of the adult population having at least a high school degree would be affected by the amount of money spent per capita by states on education. We hypothesize a positive relationship—the more money states spend on education (on a per capita basis), the higher the percentage of the adult population with at least a high school degree. Let's test this hypothesis.

Data File: **USA**
Task: **Scatterplot**
➤ Dependent Variable: **17) %HIGH SCH**
➤ Independent Variable: **41) EDUC$/CAP**

Write in the value of the correlation coefficient.   r = _~~2.40~~ .268_

Is the correlation positive or negative?
(Circle one.)   **Positive**   Negative

Is the relationship statistically significant? (Circle one.)   **Yes**   No

Do the results support the hypothesis? (Circle one.)   **Yes**   No

4. Let's examine possible relationships between the teenage motherhood rates
    in states (the percentage of all births to mothers under age 20) and drinking
    behavior.

    a. First, test the hypothesis that there is a positive relationship between the
        teenage motherhood rate and the gallons of alcoholic beverages con-
        sumed per capita.

        Data File: **USA**
        Task: **Scatterplot**
        ➤ Dependent Variable: **38) TEEN MOMS**
        ➤ Independent Variable: **52) ALCBEV/CAP**

        Write in the value of the correlation coefficient.   r = _− 0.032_

        Is the correlation positive or negative?
        (Circle one.)   Positive   **Negative**

        Is the relationship statistically significant? (Circle one.)   Yes   **No**

        Do the results support the hypothesis? (Circle one.)   Yes   **No**

    b. Now let's use a different indicator concerning alcohol consumption.
        Instead of using gallons of alcoholic *beverages* per capita, we will use gal-
        lons of *ethanol* per capita, thus focusing on how much alcohol (rather than
        alcoholic *beverages*) was actually consumed.

        Data File: **USA**
        Task: **Scatterplot**
        Dependent Variable: **38) TEEN MOMS**
        ➤ Independent Variable: **51) ETHGAL/CAP**

        Write in the value of the correlation coefficient.   r = _−.263_

        Is the correlation positive or negative?
        (Circle one.)   Positive   **Negative**

Is the relationship statistically significant? (Circle one.)   (Yes) No

Do the results support the hypothesis? (Circle one.)   Yes (No)

c. Now let's use yet another indicator concerning drinking behavior. Instead of focusing on how many gallons of alcoholic beverages were drunk per capita, let's look at the type of alcoholic beverage consumed. Specifically, let's examine the percentage of total alcoholic beverages consumed that was beer.

Before doing the analysis, what is your thought about possible relationships here? (Circle the number of your choice.)

1. There is no relationship between the teenage motherhood rate and the percentage of alcohol consumed that was beer.

2. There is a positive relationship (the higher the percentage of alcohol consumed that is beer, the higher the teenage mother rate).

3. There is a negative relationship (the higher the percentage of alcohol consumed that is beer, the lower the teenage mother rate).

> *Data File:* **USA**
> *Task:* **Scatterplot**
> *Dependent Variable:* **38) TEEN MOMS**
> ➤ *Independent Variable:* **49) %BEER**

Write in the value of the correlation coefficient.   r = __.614__

Is the correlation positive or negative?
(Circle one.)   (Positive)   Negative

Is the relationship statistically significant? (Circle one.)   (Yes)   No

Do the results support the hypothesis? (Circle one.)   (Yes)   No

d. Next, let's examine the percentage of total alcoholic beverages consumed that was wine.

Before doing the analysis, what is your thought about any possible relationship here? (Circle the number of your choice.)

1. There is no relationship between the teenage motherhood rate and the percentage of alcohol consumed that was wine.

2. There is a positive relationship (the higher the percentage of alcohol consumed that is wine, the higher the teenage mother rate).

3. There is a negative relationship (the higher the percentage of alcohol consumed that is wine, the lower the teenage mother rate).

Data File: **USA**
Task: **Scatterplot**
Dependent Variable: **38) TEEN MOMS**
➤ Independent Variable: **50) %WINE**

Write in the value of the correlation coefficient.   r = _−.593_

Is the correlation positive or negative?
(Circle one.)   Positive   (Negative)

Is the relationship statistically significant? (Circle one.)   (Yes)   No

Do the results support the hypothesis? (Circle one.)   (Yes)   No

e. Finding a correlation between two variables does not mean that one variable caused the other. Is it likely that the aggregate patterns of drinking
behavior (concerning beer and wine) examined actually affected the
teenage motherhood rate, or is it more likely that some other factor (e.g.,
differences in customs) is affecting both the teenage motherhood rate and
the type of alcoholic beverages being consumed? (Circle one.)

   1. The aggregate patterns of drinking behavior (concerning beer and
       wine) actually affected the teenage motherhood rates.

   2. Some other factor (e.g., differences in customs) is affecting both the
       teenage motherhood rate and the type of alcoholic beverages being
       consumed.

Use the MAPPING task in order to examine the relationship between the
teenage motherhood rate and the percentage of alcohol consumed that was
beer.

Data File: **USA**
➤ Task: **Mapping**
➤ Variable 1: **38) TEEN MOMS**
➤ Variable 2: **49) %BEER**

Look at each of the maps. Are there patterns here that suggest
there might be some regional patterns to both teenage motherhood rate and the percentage of alcohol consumed that is beer?
(Circle one.)   (Yes)   No

5. a. Aside from the murder rate, there are other ways in which we could have
        operationalized physical violence. Browse through the variable descrip-
        tions in the USA data file and find three other indicators of physical vio-
        lence. Write the information about them where indicated.

First indicator:
Number: _20_   Name: _KID ABUSED_
Description:

1997, NUMBER OF CHILDREN INVOLVED IN A SUBSTANTIATED
INCIDENT OF ABUSE per 100,000 POPULATION   USA

Second indicator:
Number: _40_   Name: _HATE CRIMES_
Description:

1997   reported per 100,000 POP. USA

_____

Third indicator:
Number: _102_   Name: _RAPE_
Description:

1997. FORCIBLE RAPED per 100,000, POP USA

_____

b. Now construct a hypothesis that hunting is correlated with rates of physical violence using the first indicator you listed above.

Hypothesis 1: Hunting will be (circle choice) positively / negatively correlated with ___KID ABUSE___ .
                        (first indicator)

Test this hypothesis using the SCATTERPLOT task.

Data File:	**USA**
➤ Task:	**Scatterplot**
➤ Dependent Variable:	**(the first indicator you listed)**
➤ Independent Variable:	**129) HUNTING**

Value of r?   r = _.023_

Is the relationship positive or negative? (Circle one.)   Positive   Negative

Level of significance?   Prob. = _.441_

Is this hypothesis supported? (Circle one.)   Yes   No

c. Now construct a hypothesis that hunting is correlated with rates of physical violence using the second indicator you listed previously.

Hypothesis 2: Hunting will be (circle choice) positively/~~negatively~~ correlated with ___HATE CRIME___.

(second indicator)

Test this hypothesis using the SCATTERPLOT task.

>   Data File: **USA**
>   Task: **Scatterplot**
> ➤ Dependent Variable: **(the second indicator you listed)**
> ➤ Independent Variable: **129) HUNTING**

Value of r?   r = _−.297_

Is the relationship positive or negative? (Circle one.)   Positive   Negative

Level of significance?   Prob. = ___.025___

Is this hypothesis supported? (Circle one.)   Yes   No

d. Now construct a hypothesis that hunting is correlated with rates of physical violence using the third indicator you listed previously.

Hypothesis 2: Hunting will be (circle choice) positively/~~negatively~~ correlated with ___rape___.

(third indicator)

Test this hypothesis using the SCATTERPLOT task.

>   Data File: **USA**
>   Task: **Scatterplot**
> ➤ Dependent Variable: **(the third indicator you listed)**
> ➤ Independent Variable: **129) HUNTING**

Value of r?   r = _−.170_

Is the relationship positive or negative? (Circle one.)   Positive   Negative

Level of significance?   Prob. = ___.122___

Is this hypothesis supported? (Circle one.)   Yes   No

The Research Process Using Survey Data

In Exercise 2a, we worked through an example of the research process using the USA data set. Here we will examine how to study a different research question using data on individuals. In recent years, health care reform has been the subject of a heated debate. Many individuals would like the federal government to take a more active role in the administration of health care, while others prefer to leave health care in the private sector. There are many issues in this debate, but a basic disagreement is simply over the appropriate role of the federal government in programs of this type. In general, should government be more active or less active in the day-to-day life of citizens? Those with more liberal political views favor increasing the role of government, while conservatives favor limiting its role. We would expect this difference of opinion to extend to the area of health care. So we have our first research hypothesis: Those with conservative political views are less likely to favor government intervention in health care than are those with liberal political views.

We also might expect that those who are likely to be excluded under the current health care system would favor more participation by the government. We'll test this second hypothesis too.

> ➤ *Data File:* **GSS**
> > ➤ *Task:* **Univariate**

Let's see if we can find some variables that can be used as indicators of the relevant concepts. In the variable list, click on 70) GOV MED and look at its variable description:

> SCALE ON GOVERNMENT COVERING MEDICAL COSTS: 1) I STRONGLY AGREE IT
> IS THE RESPONSIBILITY OF GOVERNMENT TO HELP; TO 5) I STRONGLY AGREE
> PEOPLE SHOULD TAKE CARE OF THEMSELVES.

This could be used as an indicator of whether individuals favor government participation in health care.

Now look at the variable description for 23) POLVIEW.[1] This question could be used as an indicator of the individual's position on the liberal/conservative

---

[1] The answer categories for POLVIEW were "recoded" by the authors so that all responses fall into three categories, rather than the original seven. You'll learn about recoding variables later in this book.

continuum. We also could use party preference, 22) POLPARTY, and how the person voted in the 1996 presidential election, 27) WHO IN 96?, as other indicators of political conservatism.

To test the second hypothesis, we need to determine who is likely to be excluded under the current health care system. Paid health care is usually not included as a benefit for low-income jobs, so 104) INCOME might be used as an indicator of those excluded. Unemployed individuals are often uncovered as well, so 3) EVER UNEMP might be used as another indicator.

The data already have been collected, so we can move to the analysis stage. First, let's look at the distribution of 70) GOV MED.

Data File: **GSS**
Task: **Univariate**
➤ Primary Variable: **70) GOV MED**
➤ View: **Pie**

GOV MED -- SCALE ON GOVERNMENT COVERING MEDICAL COSTS: 1) I strongly agree it is the responsibility of government to help; TO 5) I strongly agree people should take care of themselves. (HELPSICK)

		Freq.	%
■	1) Gov help	896	48.5
▨	2) Middle	610	33.0
▨	3) Help self	343	18.6
	TOTAL (N)	1849	100.0
	Missing	983	

The results show that 48.5 percent of the sample agreed that the government should cover medical costs. We can see that there were 983 cases with missing data—either they were not asked the question or they did not answer. Remember that missing data are excluded in all calculations, so the percentages are based only on those cases with data (1,849).

Now let's cross-tabulate views on governmental participation in health care by self-classification on the liberal-conservative dimension.

Data File: **GSS**
➤ Task: **Cross-tabulation**
➤ Row Variable: **70) GOV MED**
➤ Column Variable: **23) POLVIEW**
➤ View: **Tables**

GOV MED   by   POLVIEW
Cramer's V: 0.154 **

		POLVIEW				
		Liberal	Moderate	Conserv	Missing	TOTAL
G O V M E D	Gov help	293	316	241	46	850
	Middle	129	249	209	23	587
	Help self	52	107	174	10	333
	Missing	296	314	309	62	983
	TOTAL	474	672	624	141	1770

The cross-tabulation table shows the distribution of attitude toward government participation in health care within each category of political views. For example, 293 cases indicated that they were liberal and that they supported government health care, 316 cases were moderate and supported government health care, and so on. Missing data are omitted from all calculations, so the total for the first *row* is 850 (293 + 316 + 241), and the total for the first *column* is 474 (293 + 129 + 52).

Looking at the table, we see that, of the 474 liberals, 293 favored government health care, 129 took a middle position, and 52 did not favor government health care. Looking at the next column, we see that there were more moderates (316) than liberals (293) who favored government health care, but more moderates (107) than liberals (52) also opposed government health care. Further, the middle position was selected by more moderates (249) than liberals (129). Thus, there are more moderates than liberals in each of the categories of the government health care variable. This can happen because overall there are substantially more moderates (672) than liberals (474) in the survey.

Thus, we can't simply compare raw numbers of respondents. We must take differences in the size of population into account. To do so, we can calculate the percentage of the group in each category. Since we want to compare the views on health care across the liberal-conservative political spectrum, we need to examine *column* percentages. So, select the [Column Percentages] option.

For the rest of this discussion, we will focus on these column percentages and just ignore the raw numbers of respondents. In order to focus clearly on these column percentages, we will also ignore the column and row totals and the information concerning missing cases.

Data File: **GSS**	
Task: **Cross-tabulation**	
Row Variable: **70) GOV MED**	
Column Variable: **23) POLVIEW**	
View: **Tables**	
➤ *Display:* **Column %**	

GOV MED   by   POLVIEW
Cramer's V: 0.154 **

		POLVIEW				
		Liberal	Moderate	Conserv.	Missing	TOTAL
GOV MED	Gov help	293	316	241	46	850
		61.8%	47.0%	38.6%		48.0%
	Middle	129	249	209	23	587
		27.2%	37.1%	33.5%		33.2%
	Help self	52	107	174	10	333
		11.0%	15.9%	27.9%		18.8%
	Missing	298	314	309	62	983
	TOTAL	474	672	624	141	1770
		100.0%	100.0%	100.0%		

Column percentages allow us to compare the health care views of liberals, moderates, and conservatives more easily. Looking at the first *row* of numbers and reading from left to right, we now can see that 61.8 percent of the liberals favored government health care, compared to 47.0 percent of the moderates, and only 38.6 percent of the conservatives. Thus, there is a definite pattern in the expected direction—liberals are much more favorable toward government involvement in health care than are conservatives.

While the differences are in the predicted direction, this table is based on a sample of the population, and a sample is extremely unlikely to be exactly like the population from which it is drawn. Can we be sure that this relationship actually exists in the population? We need to look at the summary statistics to assess this.

Data File:	**GSS**
Task:	**Cross-tabulation**
Row Variable:	**70) GOV MED**
Column Variable:	**23) POLVIEW**
➤ *View:*	**Statistics (Summary)**

GOV MED   by   POLVIEW

Nominal Statistics

Chi-Square: 84.317 (DF = 4; Prob. = 0.000)					
V:	0.154	C:	0.213		
Lambda:	0.081	Lambda:	0.000	Lambda	0.033
(DV=23)		(DV=70)			

Ordinal Statistics

Gamma	0.284	Tau-b	0.183	Tau-c	0.177
s.error	0.033	s.error	0.021	s.error	0.020
Dyx	0.178	Dxy	0.189		
s.error	0.020	s.error	0.022		
Prob. =	0.000				

Look at the section of the screen labeled *nominal statistics*. Chi-square is a statistic calculated from this table. Using probability theory, statisticians can tell us how likely we are to observe a particular chi-square value by chance when there is no relationship in the population. If this probability is small enough then we can reject the hypothesis of no relationship, which means that our hypothesis that there is a relationship is supported. In this example, the probability (Prob. = 0.000) is extremely low. Consequently, we can be pretty sure that there really is a relationship between these two variables in the population. In social science, the level of .05 is used for this rejection point—if this value of chi-square would be observed fewer than 5 times in 100 when there is no relationship, then we will conclude that a relationship probably exists. This is called the *statistical significance level*. (In Chapter 4 of the text, these issues will be discussed in greater detail.)

We also can make use of another statistic on this screen, Cramer's V. The value of V tells us how strong the relationship is. This is analogous to Pearson's correlation coefficient (r), which we used when looking at scatterplots in Exercise 2a. V can range from 0 to 1. A value of 0 indicates no relationship between the two variables, and a value of 1 is the strongest possible relationship. In this example, V has a value of 0.154.

There's one extremely important way in which V is different than Pearson's r. Pearson's r tells us how well a straight line fits the data and also the direction of the relationship. V tells us nothing about the direction of the relationship, only about its strength. We must look at the table itself to describe the relationship. For example, if the values in the liberal and conservative column had been swapped, the value of V would be exactly the same. However, the table (with the swapped columns) would show that conservatives are more likely to support government health care and would consequently fail to support our hypothesis. When examining a relationship using tabular analysis, you must interpret the actual table, as well as examine the value and significance of V.

Let's try another test of the first hypothesis using a different indicator of the liberal-conservative continuum: political party preference.

Data File: **GSS**

Task: **Cross-tabulation**

Row Variable: **70) GOV MED**

➤ Column Variable: **22) POLPARTY**

➤ View: **Tables**

➤ Display: **Column %**

GOV MED by POLPARTY
Cramer's V: 0.143 **

		POLPARTY				
		Democrat	Independ	Republican	Missing	TOTAL
GOV MED	Gov help	363	342	177	14	882
		59.3%	48.6%	36.5%		49.0%
	Middle	178	240	172	20	590
		29.1%	34.1%	35.5%		32.8%
	Help self	71	122	136	14	329
		11.6%	17.3%	28.0%		18.3%
	Missing	355	366	238	24	983
	TOTAL	612	704	485	72	1801
		100.0%	100.0%	100.0%		

Be sure to use column percentaging on this table. In fact, nearly all cross-tabulation results in this workbook will use column percentaging. So get in the habit of selecting this option.

We see the same pattern here as well: Democrats are more likely than independents, and independents are more likely than Republicans, to endorse government health care. Let's check the summary statistics.

Data File: **GSS**

Task: **Cross-tabulation**

Row Variable: **70) GOV MED**

Column Variable: **22) POLPARTY**

➤ View: **Statistics (Summary)**

GOV MED by POLPARTY

Nominal Statistics

Chi-Square: 73.380 (DF = 4; Prob. = 0.000)
V:	0.143	C:	0.198		
Lambda: (DV=22)	0.032	Lambda: (DV=70)	0.000	Lambda:	0.017

Ordinal Statistics

Gamma:	0.274	Tau-b:	0.177	Tau-c:	0.170
s.error	0.032	s.error	0.021	s.error	0.020
Dyx	0.172	Dxy:	0.183		
s.error	0.020	s.error	0.021		
Prob. =	0.000				

This relationship is statistically significant at the .000 level, and V is .143. Let's test our hypothesis yet another time using another indicator (presidential voting preference) of the liberal-conservative continuum.

Data File: **GSS**

Task: **Cross-tabulation**

Row Variable: **70) GOV MED**

➤ Column Variable: **27) WHO IN 96?**

➤ View: **Tables**

➤ Display: **Column %**

GOV MED by WHO IN 96?
Cramer's V: 0.189 **

		WHO IN 96?				
		Clinton	Dole	Perot	Missing	TOTAL
GOV MED	Gov help	336	121	49	390	506
		57.0%	31.2%	39.8%		46.0%
	Middle	180	149	43	238	372
		30.6%	38.4%	35.0%		33.8%
	Help self	73	118	31	121	222
		12.4%	30.4%	25.2%		20.2%
	Missing	334	201	76	372	983
	TOTAL	569	388	123	1121	1100
		100.0%	100.0%	100.0%		

We can see that those who voted for Clinton were more likely to favor government health care than those who voted for Dole or Perot. Those who voted for Perot were more likely to favor government health care than those who voted for

Dole. Check the statistics for this relationship and you will see that it is statistically significant at the .000 level, and V is .189.

Since our hypothesis really should be limited to the Democratic and Republican candidates, we should have eliminated the Perot voters from this analysis. We can accomplish this several different ways, but the easiest method is to use the [Collapse] option that is provided with all cross-tabulation results. This feature allows us to drop the Perot supporters from the present analysis. First, with the cross-tabulation table on the screen, click on the label Perot in the Perot column. Note that this highlights the column. Now click on the [Collapse] button. Click on the option which allows you to *drop* the category—convert it to missing data for the present analysis. Click [OK] to view the updated table. Note that this has no effect at all on the column percentages for the Clinton and Bush columns. However, this procedure might change the significance level and Cramer's V. In this case, the significance level is still .000 and V is higher (.280) than it was before dropping the Perot supporters from the analysis.

All three tests of the first research hypothesis (that liberals are more supportive of government participation in health care than are conservatives) provided support for it. Let's turn to our second hypothesis—the expectation that those who are likely to be excluded under the current health care system would favor more participation by the government.

Data File:	**GSS**
Task:	**Cross-tabulation**
Row Variable:	**70) GOV MED**
➤ *Column Variable:*	**104) INCOME**
➤ *View:*	**Tables**
➤ *Display:*	**Column %**

GOV MED   by   INCOME
Cramer's V: 0.078 **

		INCOME				
		< $22,500	22.5-49.9K	$50,000 +	Missing	TOTAL
G O V M E D	Gov help	290	296	220	90	806
		54.8%	49.3%	42.6%		48.9%
	Middle	165	192	179	74	536
		31.2%	31.9%	34.6%		32.5%
	Help self	74	113	118	38	305
		14.0%	18.8%	22.8%		18.5%
	Missing	266	339	251	127	983
	TOTAL	529	601	517	329	1647
		100.0%	100.0%	100.0%		

Those in the lowest income category are the most supportive of government participation in health care, those in the next income category are somewhat less supportive, and those in the highest income category are the least supportive. The differences are not especially large, so let's check the summary statistics for this relationship. As we see, the relationship is statistically significant (Prob. = .000) and V is .078. Thus, there is a connection between income and views on government participation in health care although the relationship is not very strong. Let's try another test of this hypothesis using the question of whether the respondent has ever been unemployed.

	Data File:	**GSS**
	Task:	**Cross-tabulation**
	Row Variable:	**70) GOV MED**
➤	*Column Variable:*	**3) EVER UNEMP**
➤	*View:*	**Tables**
➤	*Display:*	**Column %**

GOV MED   by   EVER UNEMP
Cramer's V: 0.103 **

		EVER UNEMP			
		Yes	No	Missing	TOTAL
GOV MED	Gov help	313	580	3	893
		55.7%	45.6%		48.7%
	Middle	172	432	6	604
		30.6%	33.9%		32.9%
	Help self	77	261	5	338
		13.7%	20.5%		18.4%
	Missing	18	43	922	983
	TOTAL	562	1273	936	1835
		100.0%	100.0%		

The results are in the predicted direction--those who have been unemployed at some point are more favorable toward government participation in health care. When we check the statistics for this relationship, we see that it is statistically significant (Prob. = .000) and V = .103. Thus, both tests support our second hypothesis that those who are likely to be excluded under the current health care system would favor more participation by the government.

We found fairly strong support for our first hypothesis and moderate support for our second hypothesis.

Your turn.

NAME: _____

COURSE: _____

DATE: _____

EXERCISE

2b

Workbook exercises and software are copyrighted. Copying is prohibited by law.

WORKSHEET

1. Let's continue with the GSS data file and look at another relationship concerning attitudes toward government involvement in health care.

> ➤ Data File: **GSS**
> ➤ Task: **Cross-tabulation**

a. First, fill in the description of 102) OVER 50?.

Is respondent over 50? Ags

Range 1 v 2,   1) under 50 , 2) 50 and over

b. Who would you hypothesize to favor governmental assistance with health care more: people under 50 or people who are 50 or older? (Circle one.)

People under 50

✓ People 50 or older

Briefly explain your reason(s) for your hypothesis.

Some people over 50 may rely on Soc. security and medicare rather than an Employment based Health Care System (HMO etc)

c. Follow the instructions below and fill in the column percentages in the table.

> Data File: **GSS**
> Task: **Cross-tabulation**
> ➤ Row Variable: **70) GOV MED**
> ➤ Column Variable: **102) OVER 50?**
> ➤ View: **Tables**
> ➤ Display: **Column %**

	Under 50	50 & over
Gov help	_52.3_ %	_41.6_ %
Middle	_30.2_ %	_37.8_ %
Help self	_17.4_ %	_20.6_ %

Describe the relationship (the pattern in the results) in this table, focusing on the results in the first row (Gov help).

People under 50 fewer Gvernmental help.

d. Now check the summary statistics for this relationship.

> Data File: **GSS**
> Task: **Cross-tabulation**
> Row Variable: **70) GOV MED**
> Column Variable: **102) OVER 50?**
> ➤ View: **Statistics (Summary)**

$V = \underline{.104}$

e. Is the relationship statistically significant at the .05 level or better? (Circle one.)   (Yes)   No

f. Were the results in accord with what you hypothesized? (Circle one.)   Yes   (No)

If you answered **No** above, explain in one or two sentences why you think the results were different than what you expected.

People over 50 ~~get more~~ are better set to cope with health care issues than I thought they were.

2. The variable **71) MUCH GOVT** directly taps opinions about government involvement in day-to-day life.

> Data File: **GSS**
> Task: **Cross-tabulation**

a. Write the description of 71) MUCH GOVT.

Scale on how many things the government does

b. Using the same indicators of political conservatism used previously in the exercise, test the hypothesis that conservatives are more opposed to government involvement in day-to-day life. Present the results in the following tables. (Remember to use column percentaging.)

➤ *Row Variable:*   **71) MUCH GOVT**
➤ *Column Variable:* **23) POLVIEW**

	Liberal	Moderate	Conservative
Do more	145 %	153 %	108 %
Mid/2 much	318 %	496 %	508 %

V = __0.127__

Prob. = __0.000__

Row Variable: **71) MUCH GOVT**
➤ *Column Variable:* **27) WHO IN 96?**

	Clinton	Dole	Perot
Do more	172 %	45 %	16 %
Mid/2 much	403 %	337 %	108 %

V = __.217__

Prob. = __.000__

Row Variable: **71) MUCH GOVT**
➤ *Column Variable:* **22) POLPARTY**

	Democrat	Independent	Republican
Do more	190 %	167 %	68 %
Mid/2 much	406 %	521 %	403 %

V = __.158__

Prob. = __0.00__

c. In a few sentences, summarize the results of this analysis. To what extent are these indicators related to people's attitudes toward the role of government?

Democrats are more approving of Government actions

3. There's an old saying that money can't buy happiness. While that might or might not be true, let's test the hypothesis that there is a positive relationship between happiness and income. Variable 7) HAPPY? asks people whether they are very happy, pretty happy, or not too happy. Do the following analysis.

> Data File: **GSS**
> Task: **Cross-tabulation**
> ➤ Row Variable: **7) HAPPY?**
> ➤ Column Variable: **104) INCOME**
> ➤ View: **Tables**
> ➤ Display: **Column %**

a. Fill in the column percentages below. Then obtain the summary statistics and fill in Cramer's V and the significance level.

	<$22,500	$22.5 – $49.9K	$50,000 +
Not very	21.8 %	10.6 %	4.1 %
Pretty hpy	56 %	59.7 %	53.9 %
Very happy	22.2 %	29.7 %	42.0 %

V = .177

Prob. = 0.000

Based on these results, does it appear that people who have higher incomes are happier than people who have lower incomes? (Circle one.)   Yes   No

b. What other factors might affect the happiness of people? There are jokes about married people being unhappy—but there has also been some research indicating that married people are happier than unmarried people. Who do you hypothesize would be happier? (Circle one.)

Married People

Unmarried People

Let's test your hypothesis.

> Data File: **GSS**
> Task: **Cross-tabulation**
> Row Variable: **7) HAPPY?**
> ➤ Column Variable: **96) Marital**
> ➤ View: **Tables**
> ➤ Display: **Column %**

Fill in the column percentages below. Then obtain the summary statistics and fill in Cramer's V and the significance level.

	Married	Div/widowed	Nev.marry
Not very	5.3 _71_ %	19.8 _161_ %	16.5 _108_ %
Pretty hpy	51.2 _684_ %	57.0 _464_ %	65.1 _426_ %
Very happy	43.5 _582_ %	23.2 _189_ %	18.3 _120_ %

V = _.203_

Prob. = _.000_

Based on these results, does it appear that people who are married are happier than people who are not married? (Circle one.) (Yes) No

c. Some research has indicated that religious people are happier. Let's test this hypothesis that religious people are happier.

Data File:	**GSS**
Task:	**Cross-tabulation**
Row Variable:	**7) HAPPY?**
➤ Column Variable:	**84) RELPERSON**
➤ View:	**Tables**
➤ Display:	**Column %**

Fill in the column percentages below. Then obtain the summary statistics and fill in Cramer's V and the significance level.

	Slghty/not	Moderately	Very relig
Not very	15.0 %	11.5 %	13.5 %
Pretty hpy	63.3 %	54.3 %	40.2 %
Very happy	21.7 %	34.2 %	46.2 %

V = _.141_

Prob. = _0.000_

Based on these results, does it appear that people who
are religious are happier than people who are not religious?
(Circle one.)   (Yes)   No

d. What about race? Who would you hypothesize to be happier, white
Americans or African Americans? (Circle one.)

                                                                      white Americans

                                                                      African Americans

Briefly explain your answer. WITHOUT APPEARING TO BE A RACIST.

Now test your hypothesis.

Data File:	**GSS**
Task:	**Cross-tabulation**
Row Variable:	**7) HAPPY?**
➤ Column Variable:	**91) RACE**
➤ View:	**Tables**
➤ Display:	**Column %**

Fill in the column percentages below. Then obtain the summary statistics
and fill in Cramer's V and the significance level.

	White	Black	Other
Not very	10.2 %	20.2 %	19.9 %
Pretty hpy	56.1 %	57.8 %	53.3 %
Very happy	33.7 %	22.0 %	28.8 %

V = 0.093

Prob. = _____

What do you conclude about the relationship between race and happiness?

e. What about a possible relationship between happiness and the amount of time spent watching television. Would you expect this relationship to be positive or negative? (Circle one.)

Positive

~~Negative~~

Briefly explain your answer.

C'mon .

Now test your hypothesis.

> Data File: **GSS**
> Task: **Cross-tabulation**
> Row Variable: **7) HAPPY?**
> ➤ Column Variable: **2) WATCH TV**
> ➤ View: **Tables**
> ➤ Display: **Column %**

Fill in the column percentages below. Then obtain the summary statistics and fill in Cramer's V and the significance level.

	1 or less	two	3 – 4	Over 4
Not very	7.9 %	9.9 %	13.4 %	21.3 %
Pretty hpy	52.5 %	61.5 %	58.5 %	51.1 %
Very happy	39.6 %	28.6 %	28.1 %	27.1 %

V = .120

Prob. = 0.000

What do you conclude about the relationship between happiness and the amount of time spent watching television? (Circle one.)

1. There is no relationship between happiness and the amount of time spent watching television.

2. There is a positive relationship between happiness and the amount of time spent watching television—the more television people watch, the happier they are.

3. There is a negative relationship between happiness and the amount of time spent watching television—the less television people watch, the happier they are.

Briefly speculate on why the relationship between happiness and television watching exists. Is it, for example, the case that something about watching television affects the happiness level of people, or is it the case that the level of happiness of people affects how much television they watch?

4. Turning from television to radio, let's now use the NES data file to consider what kinds of people are most likely to listen to talk radio programs.

   a. The first variable to consider is 81) FAIR MEDIA, which concerns the respondent's trust of the media to be fair. What kind of relationship would you expect between frequency of listening to talk radio programs and media trust? (Circle one.)

      1. No relationship.

      2. A positive relationship—those who trust the media more will listen to talk radio programs more.

      3. A negative relationship—those who trust the media more will listen to talk radio programs less.

   Do the following analysis in order to test your hypothesis.

   > *Data File:* **NES**
   > *Task:* **Cross-tabulation**
   > *Row Variable:* **40) LSTN FREQ**
   > *Column Variable:* **81) FAIR MEDIA**
   > *View:* **Tables**
   > *Display:* **Column %**

   Fill in the column percentages below. Then obtain the summary statistics and fill in Cramer's V and the significance level.

	Nev/al nev	Some time	Most/alwys
None	_____%	_____%	_____%
Occasionally	_____%	_____%	_____%
Wkly or +	_____%	_____%	_____%

$V =$ _____

Prob. = _____

Which of the following would be the most accurate summary of these results? (Circle one.)

1. There is no relationship between media trust and frequency of listening to talk radio programs.

2. People who have higher media trust tend to listen to talk radio programs more often.

3. People who have higher media trust tend to listen to talk radio programs less often.

b. Is the frequency of listening to talk radio programs related to political views? Let's hypothesize that conservatives are more likely to listen to talk radio than moderates or liberals. Do the following analysis to test this hypothesis.

> Data File: **NES**
> Task: **Cross-tabulation**
> Row Variable: **40) LSTN FREQ**
> ➤ Column Variable: **58) LIBCON3**
> ➤ View: **Tables**
> ➤ Display: **Column %**

Fill in the column percentages below. Then obtain the summary statistics and fill in Cramer's V and the significance level.

	Liberal	Moderate	Conservative
None	_____ %	_____ %	_____ %
Occasionally	_____ %	_____ %	_____ %
Wkly or +	_____ %	_____ %	_____ %

$V =$ _____

Prob. = _____

Which of the following would be the most accurate summary of these results? (Circle one.)

1. There is no relationship between political ideology and frequency of listening to talk radio programs.

2. Conservatives tend to listen to talk radio programs more often than moderates or liberals.

3. Liberals tend to listen to talk radio programs more often than moderates or conservatives.

4. Moderates tend to listen to talk radio programs more often than liberals or conservatives.

c. Finally, let's test the hypothesis that frequency of listening to talk radio programs is positively related to political knowledge.

Data File:	**NES**
Task:	**Cross-tabulation**
Row Variable:	**40) LSTN FREQ**
➤ Column Variable:	**80) POL INFO**
➤ View:	**Tables**
➤ Display:	**Column %**

Fill in the column percentages below. Then obtain the summary statistics and fill in Cramer's V and the significance level.

	Low	Average	High
None	_____%	_____%	_____%
Occasionally	_____%	_____%	_____%
Wkly or +	_____%	_____%	_____%

V = _____

Prob. = _____

Which of the following would be the most accurate summary of these results? (Circle one.)

1. There is no relationship between political knowledge and frequency of listening to talk radio programs.

2. There is a positive relationship between political knowledge and frequency of listening to talk radio programs—people with higher knowledge listen to talk radio more often.

3. There is a negative relationship between political knowledge and frequency of listening to talk radio programs—people with higher knowledge listen to talk radio less often.

Evaluating Indicators

OVERVIEW

In this exercise, you'll learn more about measurement in social research. This will include further experience with different units of analysis (cases) and different levels of measurement. We also will delve more into reliability and validity, and see why we place more confidence in the accuracy of relationships among variables than we do in the accuracy of our measurements of the variables themselves.

BEFORE YOU BEGIN

Please make sure you have read Chapter 3 in the textbook and can answer the following review questions (you need not write any answers):

1. What does it mean to say that variables have variation within them?

2. Describe the levels of measurement: nominal, ordinal, interval, and ratio.

3. What are units of analysis, and what are some of the different kinds of units of analysis used in social research?

4. What are aggregate data, and how might misinterpretation of results based on aggregate data lead to the ecological fallacy?

5. What is reliability, and what methods can we use to assess the degree of reliability of our measurements of variables?

6. What is validity, and what are the common methods of assessing validity?

7. Compare and contrast indexes (or indices) and scales.

8. When using aggregate data, why is it often important to convert raw numbers (e.g., number of murders by state) into rates (e.g., murders per 100,000 population by state)?

In some of the physical sciences, measurement techniques are very sophisticated and precise because theories dictate exactly how various concepts should be measured. In the social sciences, we are not so fortunate, and a concept legitimately can be measured in many different ways. Consequently, our measures are much cruder and many ambiguities are introduced into the research process. In this exercise, we'll look at some of the special problems of measurement in social science.

UNITS OF ANALYSIS

Not all concepts used in the social sciences apply to individuals. Racial segregation, for example, refers to the extent to which two or more racial groups are separated from each other and, therefore, cannot be a property of individuals but only of *groups* of individuals. You can measure the racial segregation of cities, churches, schools, and so on, but you cannot measure the racial segregation of individuals.

Ideally, the unit of analysis used in the research hypothesis should be the same unit used in the research question. However, with some research questions, there are practical problems. For example, suicide is a relatively rare event. And since victims of suicide are not available to answer questions, studies of suicide frequently use aggregate units to test ideas.

In Exercise 2a, we were interested in the research question that hunting is a cause of violent behavior. We then looked at the relationship between the hunting rate and indicators of violent behaviors using states as the unit of analysis. Notice that the research question deals with the individual as the unit of analysis: Are individuals who hunt more likely to behave violently than individuals who don't hunt? However, the research hypothesis uses states as the unit of analysis and looks at the connection between the hunting rate and the murder rate. Even if the analysis at the state level had supported the research hypothesis, there would be no direct support for the underlying research question, which is trying to assess individual behavior. To do so would be to commit the ecological fallacy. When we found that the hunting rate and the murder rates were negatively correlated, we could not necessarily conclude that they were also negatively correlated at the individual level. Similarly, we could not conclude that variables correlated at the individual level necessarily would be correlated at the aggregate level.

But even studies based on less than ideal units of analysis can contribute to knowledge. For example, based on the strong negative relationship between the hunting rate and the murder rate that we observed, we can probably safely conclude that hunters are not *more* likely to commit murder.

AGGREGATES AND RATES

Sometimes, units or cases are combined to create an entirely new unit of analysis. Let's say, for example, that we wanted to examine the effect of the degree of school segregation on racial prejudice of students. One way to proceed with such a study would be to conduct surveys of students at schools that differed greatly in their racial composition, examine prejudice within those schools, and then compare the results across schools. For the first stage of the study (i.e., examining prejudice within schools), students would be the unit of analysis. But, when comparisons across the schools were made, schools would be the unit of analysis.

When aggregating information across individuals to obtain an aggregate measure, you always should transform the number into a rate for use in analysis. Let's take a look at what is meant by this.

➤ *Data File:* **USA**
➤ *Task:* **Mapping**
➤ *Variable 1:* **134) PLAYBOY#**
➤ *Variable 2:* **2) POP 98**
➤ *Views:* **Map**
➤ *Display:* **Spot Fill**

PLAYBOY# -- 1996: PLAYBOY CIRCULATION IN THOUSANDS -- COMPUTED FROM PLAYBOY CIRCULATION RATE (ABC) AND STATE POPULATION FIGURES

r = 0.990**

POP 98 -- 1998: TOTAL POPULATION IN 1000s (CENSUS)

The two maps are virtually identical. All we are seeing is that *Playboy* sells more copies in states with more people, which is hardly surprising. What we really want to know is which states have the highest *rates* of readership of *Playboy*. Hence, in order to make meaningful comparisons, we need to standardize across states on the basis of population.

Data File: **USA**
Task: **Mapping**
Variable 1: **134) PLAYBOY#**
➤ Variable 2: **114) PLAYBOY**
➤ Views: **Map**
➤ Display: **Spot Fill**

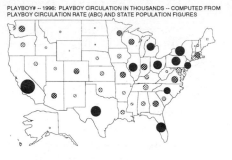

PLAYBOY# -- 1996: PLAYBOY CIRCULATION IN THOUSANDS -- COMPUTED FROM PLAYBOY CIRCULATION RATE (ABC) AND STATE POPULATION FIGURES

r = –0.256*

PLAYBOY -- 1996: PLAYBOY CIRCULATION PER 100,000 (ABC)

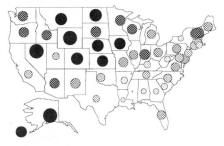

The map for PLAYBOY shows the circulation of *Playboy*, divided by the population of the state and then multiplied by 100,000. Thus, this map shows the number of *Playboy* copies sold per 100,000 population. This map is very different from the map of raw circulation numbers.

In summary, a rate is created by reducing the numbers for each unit—in this situation, each state—to a common base. Social researchers often use population as their common base, as was done in this example. This places populous states, such as California and Texas, on equal footing with states such as Wyoming and North Dakota.

LEVEL OF MEASUREMENT

Level of measurement is sometimes explained as the extent to which the categories of the variable reflect the properties of the number system. In the case of *nominal variables*, we can only sort cases into groups, such as males and females. Even if we assign a number to each group (0 for male and 1 for female), these numbers simply tell us which cases have the same gender and which cases have different genders. With *ordinal variables*, we can not only sort the cases into groups, but also order cases in terms of the property to be measured—the higher the number, the more of the property possessed by the case. *Interval variables* have yet another property of true numbers—there is an equal distance between categories. *Ratio variables* not only have equal distances between categories, but also have a rational zero point. If a case is zero on a ratio variable, this case has *none* of the property being measured.

In daily life, almost all numbers we encounter (except identification numbers such as your social security number) have all of the properties of true numbers—having zero in our checking account has a very definite meaning. Unfortunately, in social science, many of the numbers we use are not really numbers—they lack one or more of the necessary properties. Therefore, we are restricted in how we can work with them. This is why the level of measurement of a variable is important. The level reflects which properties the variable possesses.

By the way, sometimes there is confusion about whether a dichotomous variable (a variable that has only two categories, such as Favor or Oppose) is nominal or ordinal. Further, not all social researchers would actually agree on this matter. On the one hand, some social researchers feel that it is best to treat dichotomous variables as nominal. Others make a distinction between dichotomous variables that have no order and those that do have order, and this is the view that we take here. The categories of a dichotomous variable have order when one category can be said to have more of the concept being measured than the other category does. The categories of gender have no order—we can't say that males have more or less gender than females. Thus, gender is nominal. However, consider the responses to this question: *Do you favor or oppose the death penalty for those who have been convicted of murder?* The categories *favor* and *oppose* do have an order here. If we conceptualize this variable as support for capital punishment, then those who said *favor* have more support for capital punishment than those who said *oppose*.

The numbers we assign to categories of variables, referred to as *codes*, are necessary to do quantitative social science research. But just because we are assigning numbers doesn't mean the categories actually reflect all of the properties of these numbers. If we are using a race/ethnicity variable that is coded 0 for category "white," 1 for category "African American," 2 for "Asian American," and so on, we cannot use these numbers to order cases from less to more on our race variable. We cannot say, for example, that zero represents the absence of race. We are merely using numbers to show whether cases are in the same category or in different categories.

When we analyze data, we use many different techniques, or statistics, for summarizing the observed data. Each of these techniques makes assumptions about the numeric properties the variable possesses (i.e., its level of measurement). A frequency distribution is the simplest type of summary statistic and can be used with any level of measurement. Let's examine the frequency distribution for a nominal variable.

➤ *Data File:* **GSS**
➤ *Task:* **Univariate**
➤ *Primary Variable:* **74) RELIGION**
➤ *View:* **Statistics (Summary)**

RELIGION -- What is your religious preference? Is it Protestant, Catholic, Jewish, some other religion, or no religion? (RELIG)

Mean:	1.779	Std.Dev.:	1.101	N:	2703
Median:	1.000	Variance:	1.213	Missing:	129
99% confidence interval +/- mean: 1.724 to 1.833					
95% confidence interval +/- mean: 1.737 to 1.820					

Category	Freq.	%	Cum.%	Z-Score
1) Protestant	1524	56.4	56.4	-0.707
2) Catholic	705	26.1	82.5	0.201
3) Jewish	50	1.8	84.3	1.109
4) None	396	14.7	99.0	2.017
5) Other	28	1.0	100.0	2.925

Notice that numbers have been assigned to the categories: Protestants are coded 1, Catholics are coded 2, and so on. This is a nominal variable, since cases can only be categorized, not ordered, by religious affiliation.

The category Protestant has more cases than any other category. Again, do not confuse the category *code*, 1, with the number of cases in the category, 1524. Now let's examine the frequency distribution for an ordinal variable.

Data File: **GSS**
Task: **Univariate**
➤ *Primary Variable:* **126) ATTEND!**
➤ *View:* **Statistics (Summary)**

#SIBS! -- How many brothers and sisters did you have? Please count those born alive, but no longer living, as well as those alive now. Also include stepbrothers and stepsisters, and children adopted by your parents. (SIBS)

Mean:	3.980	Std.Dev.:	3.585	N:	2815
Median:	3.000	Variance:	12.853	Missing:	17
99% confidence interval +/- mean: 3.806 to 4.154					
95% confidence interval +/- mean: 3.847 to 4.112					

Value	Freq.	%	Cum.%	Z-Score
0	105	3.7	3.7	-1.110
1	493	17.5	21.2	-0.831
2	558	19.8	41.1	-0.552
3	483	17.2	58.2	-0.273
4	350	12.4	70.7	0.006
5	207	7.4	78.0	0.285
6	157	5.6	83.6	0.564
7	118	4.2	87.8	0.842
8	84	3.0	90.8	1.121
9	68	2.4	93.2	1.400
10	43	1.5	94.7	1.679

The categories of ATTEND! are also coded with numbers: Those who never attend religious services are coded 0, those who attend only once a year or less often are coded 1, and so on. These codes allow us not only to group cases as similar or different, but also to *order* cases. So this variable is an ordinal variable. By convention, the lowest category is usually coded 1 (or 0), and the code is incremented by 1 for each additional category.

Look at the column labeled "Cum. %." The numbers in this column represent the percentage of cases at or below that category. For example, we see that 59.1 percent attend religious services once a month or less often. If we couldn't order the categories, this summary measure wouldn't be valid. Looking at the previous example, we could not have said that 82.5 percent of the sample are Catholic or lower—no one would know how to interpret the statement. So cumulative percentages should be used only with ordinal or higher levels of measurement.

Percentile is another word for cumulative percentage. The 90th percentile of a group on a particular variable is the category at or below which 90 percent of the group scored. The *median* is the 50th percentile. The median is a commonly used measure because it splits a sample in half—50 percent of the sample are at or below the median value. Next, let's look at a ratio variable.

Data File: **GSS**
Task: **Univariate**
➤ Primary Variable: **117) #SIBS!**
➤ View: **Statistics (Summary)**

#SIBS! -- How many brothers and sisters did you have? Please count those born alive, but no longer living, as well as those alive now. Also include stepbrothers and stepsisters, and children adopted by your parents. (SIBS)

Mean:	3.980	Std.Dev	3.585	N	2815
Median:	3.000	Variance:	12.853	Missing:	17

99% confidence interval +/- mean: 3.806 to 4.154
95% confidence interval +/- mean: 3.847 to 4.112

Value	Freq.	%	Cum.%	Z-Score
0	105	3.7	3.7	-1.110
1	493	17.5	21.2	-0.831
2	558	19.8	41.1	-0.552
3	483	17.2	58.2	-0.273
4	350	12.4	70.7	0.006
5	207	7.4	78.0	0.285
6	157	5.6	83.6	0.564
7	118	4.2	87.8	0.842
8	84	3.0	90.8	1.121
9	68	2.4	93.2	1.400
10	43	1.5	94.7	1.679

Since the categories of #SIBS! are ordered, the median is meaningful. In the statistics listed on the screen, you can see that the median number of siblings for this sample was 3.00. This means that half of the sample had 3 sibs or fewer and the other half had more.

Because this measure possesses equal distance between categories (a requirement of interval and ratio variables), we can add category values. Consequently, we can find the *average*, or *mean*, value of the variable. The average is found by adding the category values over all cases and dividing by the number of cases. Looking at the statistics on the screen, we can see that the average number of siblings for this sample was 3.98.

Both the median and the mean tell us something about the location of the middle of the distribution. Such statistics are broadly referred to as *measures of central tendency*.

Statistics that summarize the relationship *between* variables also make assumptions about the level of measurement of the variables involved. For example, in Exercise 2b, we saw that Cramer's V provides information on a cross-tabulation that is similar to the information provided by Pearson's r on a scatterplot. The main difference between these statistics and their interpretation is the level of measurement of the variables. Scatterplots and Pearson's r assume that the variables are measured at the interval or ratio level, while Cramer's V only requires nominal level of measurement. Let's look at the statistics for a cross-tabulation table.

Data File: **GSS**
➤ Task: **Cross-tabulation**
➤ Row Variable: **85) PRAYFREQ**
➤ Column Variable: **101) AGE**
➤ View: **Tables**
➤ Display: **Column %**

PRAYFREQ   by   AGE
Cramer's V: 0.186 **

	AGE				
	18-29	30-39	40-49	50-64	65 &over
< Daily	168	159	124	122	72
	60.2%	47.0%	42.9%	43.6%	30.0%
Daily/more	111	179	165	158	168
	39.8%	53.0%	57.1%	56.4%	70.0%
Missing	208	349	291	265	239
TOTAL	279	338	289	280	240
	100.0%	100.0%	100.0%	100.0%	100.0%

Row percentages, column percentages, and total percentages can be used here, but we'll use only column percentages in the exercises. Notice that there is a pattern in these results—there are differences in the frequency of praying among different age groups. Now we need to examine the statistics for this relationship between frequency of praying and age.

Data File: **GSS**
Task: **Cross-tabulation**
Row Variable: **85) PRAYFREQ**
Column Variable: **101) AGE**
➤ View: **Statistics (Summary)**

PRAYFREQ   by   AGE

Nominal Statistics

Chi-Square: 49.150	(DF =	4; Prob. = 0.000)			
V:	0.186	C:	0.183		
Lambda: (DV=101)	0.008	Lambda: (DV=85)	0.088	Lambda:	0.038

Ordinal Statistics

Gamma:	0.243	Tau-b:	0.154	Tau-c:	0.193
s.error	0.037	s.error	0.023	s.error	0.029
Dyx:	0.121	Dxy:	0.195		
s.error	0.018	s.error	0.029		
Prob. =	0.000				

Notice the "Nominal Statistics" heading at the top of the screen. All the statistics in this group are appropriate for use with nominal variables. Remember that, while Cramer's V indicates the *strength* of the relationship, it does not tell us the direction or nature of the relationship. To describe a nominal relationship, you must look at the table itself.

In the middle of the screen, you will see another set of statistics (gamma, tau-b, tau-c, dyx, and dxy) that are for ordinal variables. Since ordinal variables (and variables with higher levels of measurement) have a natural direction or order to their categories, the relationship between variables can be either positive or negative—as one variable increases, the other variable may either increase or decrease. Hence, the statistics under this heading describe both the strength and the direction of the relationship, just as the correlation coefficient does.

Your instructor may wish you to report the value and significance of gamma or Kendall's tau when both variables are ordinal. Gamma or Kendall's tau may range from –1 through 0 to +1, where the sign shows the direction of the relationship and the value indicates the strength. The significance of all the ordinal statistics is shown at the bottom (Prob. = 0.000).

Be alert to the fact that MicroCase will provide values for all the statistics regardless of the level of measurement—this is true of all leading statistical packages. Consequently, in any analysis *you* must determine which statistics are appropriate. For example, it would be inappropriate to report the value of gamma or Kendall's tau for nominal variables.

As pointed out in the text, few social science variables based on individuals can be considered truly interval or ratio. For example, while years served in prison may be a ratio measure of severity of sentence, years spent in school is probably not a ratio measure of education. Unfortunately, the most powerful statistical techniques assume at least an interval level of measurement. But the use of these techniques on ordinal-level variables does not seem to generate misleading conclusions in most applications, and statisticians are even finding ways to use these techniques with variables measured at the nominal level.

In this course, unless your instructor requests otherwise, we will assume that correlation and other related techniques can be used with ordinal or higher levels of measurement. You'll learn more about potential problems with this assumption when you take a statistics course.

RELIABILITY AND VALIDITY

Based on the discussion in the text, you may have noticed that we always assess reliability by looking at the relationship *between* two variables:

- Inter-rater reliability—relationship between ratings by different individuals

- Test-retest reliability—relationship between scores on the same test at different times

- Alternate forms reliability—relationship between scores on two different forms of the same test

- Split halves reliability—relationship between scores on two different forms administered at the same time

- Internal consistency reliability—relationships among items used in an index

Let's look at Cronbach's alpha—a measure of reliability based on internal consistency.

Data File: **GSS**
➤ Task: **Correlation**

First, look at the variable description for 33) FREESPEAK. This support for freedom of speech index is based on answers to five questions (variables 28–32) asking whether the respondent would allow a public speech by a person opposed to religion and churches, a racist, a communist, a militarist, and a homosexual.

Is this measure reliable in terms of internal consistency? It is possible that some of these questions are reliable indicators while others are not. While social scientists usually focus more on validity than on reliability, let's look at one common measure of reliability that could be used here: Cronbach's alpha coefficient. Alpha is used to measure internal consistency—based on the extent to which the items are correlated with one another—especially when a researcher is developing an index to measure some concept.

Cronbach's alpha varies from 0 (completely unreliable) to 1.0 (perfectly reliable in terms of internal consistency). If alpha is 0.7 or higher, then the index is considered to be reliable. Let's check the reliability for the support for freedom of speech index.

          Data File: **GSS**
          Task: **Correlation**
➤ *Select Variables:* **28) ATHSPK**
          29) RACSPK
          30) COMSPK
          31) MILSPK
          32) GAYSPK

Correlation Coefficients
N: 1730   Missing: 1102
Cronbach's alpha: 0.829
LISTWISE deletion (1-tailed test)   Significance Levels: ** = .01, * = .05

	ATHSPK	RACSPK	COMSPK	MILSPK	GAYSPK
ATHSPK	1.000	0.511 **	0.584 **	0.510 **	0.517 **
RACSPK	0.511 **	1.000	0.489 **	0.461 **	0.372 **
COMSPK	0.584 **	0.489 **	1.000	0.595 **	0.481 **
MILSPK	0.510 **	0.461 **	0.595 **	1.000	0.430 **
GAYSPK	0.517 **	0.372 **	0.481 **	0.430 **	1.000

At the top of the screen, you can see that Cronbach's alpha is 0.829 and hence our index is reliable. You can see that this high value results from the high correlations among the component variables.

Similarly, all tests of validity, except face validity, are also based on examining the relationship between variables. For example, in the GSS data set, individuals are asked how often they attend religious services and also how strong their religious preference is. If the measure of religiosity is valid, we would expect that those who attend religious services more often would have stronger religious preferences. So we could look at the relationship between these variables to test the validity of the religiosity measure.

          Data File: **GSS**
➤   *Task:* **Cross-tabulation**
➤   *Row Variable:* **83) RELPREF S**
➤ *Column Variable:* **88) ATTEND**
➤   *View:* **Tables**
➤   *Display:* **Column %**

RELPREF S   by   ATTEND
Cramer's V: 0.507 **

		Not often	Often	Missing	TOTAL
	Less	938	299	6	1237
		78.7%	28.2%		54.9%
	Strong	254	763	5	1017
		21.3%	71.8%		45.1%
	Missing	457	77	33	567
	TOTAL	1192	1062	44	2254
		100.0%	100.0%		

We can see from these column percentages that those who attend religious services often claim to have stronger religious preferences than those who don't. (Further, a check of the statistics shows that Cramer's V = .507 and Prob. = .000.) Hence, we have more confidence in the validity of our measure of religiosity.

WHY RELATIONSHIPS?

This focus on relationships is also a result of the crudeness of our measures. Very few social science variables have categories that allow us to give meaningful interpretations to the distribution of a single variable. Generally, only variables with "natural" categories, such as sex, race, and political party affiliation, or variables measured at the ratio level, such as age and income, provide meaningful distributions. For example, the statement that a group is 70 percent female or the statement that the average age of the group is 63.5 can be unambiguously interpreted. However, distributions of single variables that have a natural continuum but no meaningful zero point, such as degree of belief in God, attitude toward abortion, or degree of confidence in the government, cannot be interpreted easily.

For example, the statement that 60 percent of the population has a great deal of confidence in the government has little meaning. How much is a "great deal"? We don't know whether this is higher or lower than the confidence held for other groups, such as big business and labor unions, or even whether confidence in government is higher or lower than it was last year or 10 years ago.

Analysis of individual variables is also problematic since relatively minor changes in the wording of questions may have sizable effects on the distribution of responses. In the GSS data file, some questions have alternate wordings, whereby half the sample (randomly selected) was asked one version of a question and half was asked the second version. This was a methodological study to determine how much changes in question wordings would affect the responses. Let's examine one example of this.

	Data File:	**GSS**
	➤ Task:	**Univariate**
➤	Primary Variable:	**34) WELFARE $**
	➤ View:	**Pie**

WELFARE $ -- Spending on welfare (NATFARE)

		Freq.	%
■	1) Too much	598	45.4
▨	2) Right	498	37.8
▩	3) Too little	221	16.8
	TOTAL (N)	1317	100.0
	Missing	1515	

This survey question asked for respondents' opinions on government spending on "welfare." Examine the percentage distributions for a moment. Then look at the alternate question, 35) WELFARE $2.

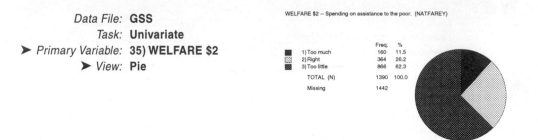

Data File:	**GSS**
Task:	**Univariate**
➤ Primary Variable:	**35) WELFARE $2**
➤ View:	**Pie**

WELFARE $2 -- Spending on assistance to the poor. (NATFAREY)

		Freq.	%
■	1) Too much	160	11.5
▨	2) Right	364	26.2
▩	3) Too little	866	62.3
	TOTAL (N)	1390	100.0
	Missing	1442	

What may seem like slight changes in question wording had a huge effect on the percentage who selected each category. When asked about spending on welfare, only 16.8 percent thought the government was spending too little, but when asked about assistance to the poor, 62.3 percent thought too little was being spent.

Before seeing the data, almost any researcher would have accepted these questions as interchangeable indicators of the respondent's attitude toward helping the poor. After we see the data, we can, of course, speculate on how these questions must be tapping somewhat different attitudes, but such *ex post facto* analysis will not help us design indicators for another concept. If such minor changes in wording will make such a big change in the resulting distribution, how can social scientists ever develop adequate indicators of concepts?

Fortunately, as social scientists, we are primarily interested in the *relationship* between variables. It turns out that relationships between variables are usually much less sensitive than distributions of simple variables to changes in measurement techniques. Despite their limitations as descriptions, both of these question wordings are useful for comparing across groups. For example, we could use this variable to test the hypothesis that lower-income respondents favor spending to help the poor more than upper-income respondents do. First, let's use the cross-tabulation task and select WELFARE $ as the row variable and INCOME as the column variable.

Data File:	**GSS**
➤ Task:	**Cross-tabulation**
➤ Row Variable:	**34) WELFARE $**
➤ Column Variable:	**104) INCOME**
➤ View:	**Tables**
➤ Display:	**Column %**

WELFARE $ by   INCOME
Cramer's V: 0.097 **

		INCOME				
		< $22,500	22.5-49.9K	$50,000 +	Missing	TOTAL
WELFARE $	Too much	134	221	175	68	530
		35.8%	49.3%	49.3%		45.0%
	Right	156	158	132	52	446
		41.7%	35.3%	37.2%		37.9%
	Too little	84	69	48	20	201
		22.5%	15.4%	13.5%		17.1%
	Missing	421	492	413	189	1515
	TOTAL	374	448	355	329	1177
		100.0%	100.0%	100.0%		

Compare the percentages of people in different income groups who think too little is being spent. As the income levels increase, the percentage of those who think too little is being spent on welfare decreases. Also look at the summary statistics and note that Cramer's V = .097 and the significance level = .000. Now make the alternate question using WELFARE $2 as the row variable and INCOME as the column variable.

Data File:	**GSS**
Task:	**Cross-tabulation**
➤ *Row Variable:*	**35) WELFARE $2**
➤ *Column Variable:*	**104) INCOME**
➤ *View:*	**Tables**
➤ *Display:*	**Column %**

WELFARE $2   by   INCOME
Cramer's V: 0.074 **

		INCOME				
		< $22,500	22.5-49.9K	$50,000 +	Missing	TOTAL
WELFARE $2	Too much	43	49	52	16	144
		10.9%	10.7%	13.6%		11.7%
	Right	82	118	116	48	316
		20.8%	25.8%	30.3%		25.6%
	Too little	270	290	215	91	775
		68.4%	63.5%	56.1%		62.8%
	Missing	400	483	385	174	1442
	TOTAL	395	457	383	329	1235
		100.0%	100.0%	100.0%		

Compare the two groups again and then look at the summary statistics. For this relationship, Cramer's V = .074 and the significance level = .009. Note that, regardless of the wording difference, the *relationship* between income level and attitude toward spending to help the poor is the same. Our hypothesis would be supported with either variable: Lower-income respondents are more likely than upper-income respondents to think too little is being spent on helping the poor, and the results are statistically significant.

Studies that examine changes over time are also looking at relationships—time is one of the variables. For example, if we compare the distribution of attitudes toward spending on the poor in 1972 with the distribution of this attitude in 1998, we are looking at the effect of "time" on this attitude. Of course, for such a comparison to be valid, the same measure of attitude toward spending on the poor must be used each time.

Your turn.

NAME: _____

COURSE: _____

DATE: _____

Workbook exercises and software are copyrighted. Copying is prohibited by law.

EXERCISE

3

WORKSHEET

1. Let's begin by looking at variables in the NES data file.

    ➤ *Data File:* **NES**
        ➤ *Task:* **Cross-tabulation**

    a. What is the variable description for 106) TAKE ADVAN?

    Do you think most people would try to take advantage of you if they got the chance or would they try to be fair?

    b. What concept do you think this variable is measuring?

    c. What is the variable description for 107) TRUST?

    Generally speaking, would you say that most people can be trusted, or that you can't be too careful in dealing with people

d. Let's cross-tabulate these two variables. Fill in the column percentages in the table that follows.

Data File:	**NES**
Task:	**Cross-tabulation**
➤ Row Variable:	**106) TAKE ADVAN**
➤ Column Variable:	**107) TRUST?**
➤ View:	**Tables**
➤ Display:	**Column %**

	Be careful	Can trust
Take advn	61.3 ~~39~~ %	15.6 %
Be fair	38.7 %	84.4 %

e. Let us assume for the moment that 106) TAKE ADVAN is a fairly valid measure of the concept that you specified in question 1b previously. Based on this table, how valid is variable 107) TRUST? in measuring this same concept? (Circle one.)

Not very valid

Somewhat valid

Very valid

Explain your answer.

2. Read the variable descriptions for the following variables in the NES data file and indicate the level of measurement of each by circling *nominal*, *ordinal*, or *interval/ratio*. **Note:** Be sure to examine the categories used with each variable before you decide which level of measurement is involved.

1) AGE	Nominal	Ordinal	Interval/ratio
2) AGE CATEGR	Nominal	Ordinal	Interval/ratio
7) REGION	Nominal	Ordinal	Interval/ratio
18) REL FREQ	Nominal	Ordinal	Interval/ratio
19) RELPREF	Nominal	Ordinal	Interval/ratio
50) APPRV REP	Nominal	Ordinal	Interval/ratio
70) WOMEN EQL	Nominal	Ordinal	Interval/ratio
79) POLITKNOW	Nominal	Ordinal	Interval/ratio

3. Different kinds of research require different kinds of units of analysis (cases). For each of the following research questions, circle **Yes** if it would be appropriate to use individuals as the units of analysis (cases) or **No** if it would not be appropriate.

a. Is the divorce rate in coastal states of the United States higher than the divorce rate in noncoastal states? (Circle one.)   (Yes) (No)

b. Do the attitudes of people on the abortion issue have any connection with their attitudes on the issue of capital punishment? (Circle one.)   (Yes) (No)

c. Does the crime rate increase when unemployment increases? (Circle one.)   (Yes) (No)

d. To what extent is the voting behavior of the U.S. representatives on bills in Congress related to the number of terms they have served in Congress? (Circle one.)   (Yes) No

e. Do states that have legalized gambling have lower taxes? (Circle one.)   Yes (No)

f. To what extent are the religious views of people related to their views on political issues? (Circle one.)   (Yes) No

4. In the Canadian General Social Survey conducted by Statistics Canada, the following question asks respondents about church attendance:

Other than on special occasions, such as weddings, funerals or baptisms, how often did you attend services or meetings connected with your religion in the last 12 months? Was it . . .

> *At least once a week?*
> *At least once a month?*
> *A few times a year?*
> *At least once a year?*
> *Not at all*

a. Using the GSS data set, record the variable description and answer categories for 88) ATTEND.

Description:

How often do you attend religious services

Categories:

1) Not often   2) often

b. What concept is being measured by each of these two variables?

c. Which of these church attendance items is the better indicator? Why?

d. Record the description and answer categories of 126) ATTEND!.

Description:

Same

Categories:

0) Never,   1) <1 a year   2) 1 n 2 yr   3) several/year
4) 1 a month   5) 2-3 month

e. Let's compare the strengths and weaknesses of
   this indicator with the one used by Statistics Canada.
   Which indicator allows for the measurement of greater
   variation? (Circle one.)   ⃝126) ATTEND!

                                                        Statistics Canada question

f. Which indicator appears to be better at measuring
   the frequency with which people *ordinarily* attend
   religious services? (Circle one.)   126) ATTEND!

                                                        ⃝Statistics Canada question

5. Find the percentages of survey respondents who would allow an abortion in
   each of the situations specified in variables 49–54, and write these percentages in the blanks provided.

> *Data File:* **GSS**
> *Task:* **Univariate**
> *Primary Variable:* **49) ABORT DEF (Repeat with 50) ABORT WANT,**
> **51) ABORT HLTH . . .)**
> *View:* **Pie**

49) ABORT DEF ~~42.3~~ 77.9 %

50) ABORT WANT 42.3 %

51) ABORT HLTH 87.5 %

52) ABORT POOR 44.2 %

53) ABORT RAPE 79.4 %

54) ABORT SING 42.3 %

Now go back to the NES data file and obtain univariate statistics for variable 33) ABORTION.

> *Data File:* **NES**
> *Task:* **Univariate**
> *Primary Variable:* **33) ABORTION**
> *View:* **Pie**

The variable description for 33) ABORTION is given below along with blanks next to each of the four response categories. Write in the percentages who selected each of the options.

There has been some discussion about abortion during recent years. Which one of the opinions on this page best agrees with your view?

12.3% By law, abortion should never be permitted.

29.1 % The law should permit abortion ONLY in case of rape, incest, or when the woman's life is in danger.

16.7 % The law should permit abortion for reasons OTHER THAN rape, incest, or danger to the woman's life, but only after the need for the abortion has been clearly established.

41.9 % By law, a woman should always be able to obtain an abortion as a matter of personal choice.

Examine the results for this NES indicator of abortion attitudes, and then take another look at the results for the six GSS indicators of abortion attitudes. On the basis of the results for these seven indicators, what would be the appropriate response to the question "Where do Americans stand on the abortion issue?" Explain your answer.

6. Let's return to the GSS file. In addition to 34) WELFARE $ and 35) WELFARE $2 discussed in this exercise, alternate wordings of several other variables were included in the methodological study for GSS. Let's examine some other examples of alternate wordings.

    a. Provide the variable description and distribution for the following variable:

      ➤ Data File: **GSS**
      ➤ Task: **Univariate**
      ➤ Primary Variable: **40) ENVIRON $**
      ➤ View: **Pie**

    Description:

    _____

    _____

    In the following table, fill in the appropriate frequency and percentage distributions.

	Frequency	Percentage
Too much	_____	_____%
Right	_____	_____%
Too little	_____	_____%

b. Provide the variable description and distribution for the following variable.

> Data File: **GSS**
> Task: **Univariate**
> ➤ Primary Variable: **41) ENVIRON $2**
> ➤ View: **Pie**

Description:

_____

Fill in the following table.

	Frequency	Percentage
Too much	_____	_____ %
Right	_____	_____ %
Too little	_____	_____ %

c. Compare the results for these two variables. Do you think it is legitimate to consider these two variables as indicators of the same concept? (Circle one.)   Yes   No

d. If you answered **Yes**, provide a definition of the concept. If you answered **No**, provide definitions of the two concepts being measured.

e. Now examine the relationship between 23) POLVIEW and the first of these environmental variables. The hypothesis is: Liberals are more likely to think too little is being spent on the environment than are conservatives.

> Data File: **GSS**
> ➤ Task: **Cross-tabulation**
> ➤ Row Variable: **40) ENVIRON $**
> ➤ Column Variable: **23) POLVIEW**
> ➤ View: **Tables**
> ➤ Display: **Column %**

Fill in the column percentages in the following table, and then fill in the Cramer's V and the significance level for the relationship.

	Liberal	Moderate	Conservative
Too much	_____%	_____%	_____%
Right	_____%	_____%	_____%
Too little	_____%	_____%	_____%

V = _____

Prob. = _____

Do the results support the hypothesis? (Circle one.)   Yes   No

f. Now examine the relationship between 23) POLVIEW and the second of these environmental variables. The hypothesis is still: Liberals are more likely to think too little is being spent on the environment than are conservatives.

> Data File: **GSS**
> Task: **Cross-tabulation**
> ➤ Row Variable: **41) ENVIRON $2**
> ➤ Column Variable: **23) POLVIEW**
> ➤ View: **Tables**
> ➤ Display: **Column %**

Fill in the column percentages in the following table, and then fill in the Cramer's V and the significance level for the relationship.

	Liberal	Moderate	Conservative
Too much	_____%	_____%	_____%
Right	_____%	_____%	_____%
Too little	_____%	_____%	_____%

V = _____

Prob. = _____

Do the results support the hypothesis? (Circle one.)   Yes   No

g. In testing this hypothesis (liberals are much more likely to think too little is being spent on the environment than conservatives are), did it make a difference in the results which indicator of environmental spending attitudes (ENVIRON $ or ENVIRON $2) you selected? (Circle one.)   Yes   No

Explain your answer below.

7. ABORT INDX is a support-for-abortion index consisting of the number of **Yes** answers respondents gave to six abortion questions (49–54). Let's examine Cronbach's alpha reliability coefficient for this index. Remember that we want alpha to be at least .7 in order to decide that the variables constituting an index provide a reliable measure.

> | | |
> |---:|:---|
> | *Data File:* | **GSS** |
> | ➤ *Task:* | **Correlation** |
> | ➤ *Select Variables:* | **49) ABORT DEF** |
> | | **50) ABORT WANT** |
> | | **51) ABORT HLTH** |
> | | **52) ABORT POOR** |
> | | **53) ABORT RAPE** |
> | | **54) ABORT SING** |

a. What is the alpha for this set of variables?   Alpha = _____

b. On the basis of the value of alpha, would you conclude that the support-for-abortion index (ABORT INDX) is reliable or unreliable? (Circle one.)   Reliable   Unreliable

In the area of abortion attitudes, we might want to distinguish two clusters of attitudes: one set that broadly concerns emergencies and another set that is concerned with less threatening matters. What happens to the degree of reliability if we split these six variables into two sets of three? Let's consider the three abortion questions that concern emergencies first.

> | | |
> |---:|:---|
> | *Data File:* | **GSS** |
> | *Task:* | **Correlation** |
> | ➤ *Select Variables:* | **49) ABORT DEF** |
> | | **51) ABORT HLTH** |
> | | **53) ABORT RAPE** |

c. What is the alpha for this set of variables?   Alpha = _____

d. On the basis of the value of alpha, would you
   conclude that this support-for-abortion index
   is reliable or unreliable? (Circle one.)   Reliable   Unreliable

Now let's look at the correlations for the other three abortion questions.

> Data File: **GSS**
>   Task: **Correlation**
> ➤ Select Variables: **50) ABORT WANT**
>   **52) ABORT POOR**
>   **54) ABORT SING**

e. What is the alpha for this set of variables?   Alpha = _____

f. On the basis of the value of alpha, would you
   conclude that this support-for-abortion index
   is reliable or unreliable? (Circle one.)   Reliable   Unreliable

Selecting Cases

OVERVIEW

In this exercise, you will learn more about problems that can cause bias in samples, and you will also see that some sampling problems probably do not seriously distort the accuracy of the results.

BEFORE YOU BEGIN

Please make sure you have read Chapter 4 in the textbook and can answer the following review questions (you need not write any answers):

1. What is the difference between a census and a sample?

2. What is the difference between a parameter and a statistic?

3. If the confidence interval for a sample is plus or minus 3 percentage points and the confidence level is 95 percent, what does this mean in terms of the results?

4. Suppose you have a list of all students in your college and you want to take a sample of them. How would you go about selecting (a) a simple random sample and (b) a systematic random sample?

5. Describe the general process by which you would select a probability sample of people in a city for a telephone survey.

6. Describe how each of the following sources of bias can distort random samples: nonresponse bias, selective availability bias, areal bias.

7. What is a SLOPS, and how trustworthy are the results of such surveys?

The first step in selecting cases is to define the relevant population. When one is primarily interested in *describing* a particular group of units, determining the population is straightforward. A magazine, for example, might be interested in the opinions of its current subscribers—the appropriate population is individuals who subscribe to the magazine. Or a college might want to determine how its students budget their time—students enrolled at the school would be the population of interest. A city might want to evaluate the utilization of recreational facilities—the available facilities might be used as the cases.

When research shifts to *testing ideas* rather than describing groups, defining the appropriate population is not so easy. For example, a researcher testing a theory about the effect of organizational structure on decision making would probably want to define the relevant population as all organizations over all time. Similarly, someone interested in the effect of new and contradictory information on attitude change might want to generalize the results to all humans over all time. Clearly, however, there is no way to take a census of either of these populations or even to draw scientific samples. Consequently, researchers settle for working with a subset of the population of interest.

For example, the organization researcher might define the population as all business firms within a particular geographic area at the current time. The attitude change researcher might define the population as individuals over age 18 currently residing in the United States. Other researchers testing these same theories might choose quite different populations. In fact, one type of replication research is to test the same hypotheses in a different population. Thus, a researcher might replicate the organizational study by using national charitable organizations as the relevant population.

Having defined the appropriate population, the next step is to determine how cases are to be selected from this population. For many research questions, using a census of the units is more appropriate than selecting a sample. If an instructor were interested in the opinions and attitudes of students in a particular class, he or she should simply take a census of the class—that is, include all members in the study. There is no reason to select a sample, since the entire population is accessible and relatively small—why introduce unnecessary sources of inaccuracy? If one is studying state governments, governments of all 50 states would be included, unless the data collection process was quite complicated. In short, samples are used only when taking a census would be more difficult.

If a sample is to be used, researchers must determine how to select the sample. If we wish to generalize the results from the sample to the population, then we must use a probability (or random) sample. With such samples, we can determine the exact probability that a particular case in the population is included in the sample. Using probability theory, we can then generalize from the observed sample to the population. This process of inferring from a sample to a population will be covered in detail when you take a statistics course. For now, you can simply assume that, when a proper sample is selected, the sample statistic provides the best estimate of the population parameter. For example, if the proportion of females in the sample is .63 (sample statistic), then our best estimate of the proportion of females in the population is .63 (population parameter).

An important property of probability samples is that two or more random samples of the same population can be combined and this combined sample is also a probability sample. This can be very useful. Suppose, for example, you find that a substantial number of cases in your sample are no longer members of the selected population—perhaps they have moved to an area not included in your sampling frame. As a result, your sample will have many fewer cases than you had planned. You can replace these cases by simply drawing another sample, including sufficient cases to complete the original sample, and combine these data with your first sample (you would, of course, want to drop any duplicates that appear).

The problem with an *improperly* selected sample is that estimates of the population parameters are almost certain to be incorrect. Let's look at an example. Suppose you administered a survey to those attending Sunday services at several churches. Let's take a look at attendance at religious services.

➤ *Data File:* **GSS**
➤ *Task:* **Univariate**
➤ *Primary Variable:* **88) ATTEND**
➤ *View:* **Pie**

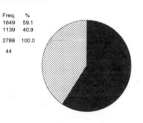

ATTEND -- How often do you attend religious services? (ATTEND) (The Often category includes 2-3 times a month or more often.

		Freq.	%
■	1) Not often	1649	59.1
▨	2) Often	1139	40.9
	TOTAL (N)	2788	100.0
	Missing	44	

These results indicate that unless some unusual Sunday is selected, we would expect less than half the population to be in church. But the real problem is that those who are in church will differ in many ways from those not in church. Let's look at some of these differences. First, let's find out what the percentages of males and females are in the total sample.

Data File: **GSS**
Task: **Univariate**
➤ *Primary Variable:* **89) SEX**
➤ *View:* **Pie**

SEX -- Respondent's sex (SEX)

		Freq.	%
■	0) Male	1232	43.5
▨	1) Female	1600	56.5
	TOTAL (N)	2832	100.0

The percentage of males in the total sample is 43.5. Now let's look at the sex distribution of church attendees.

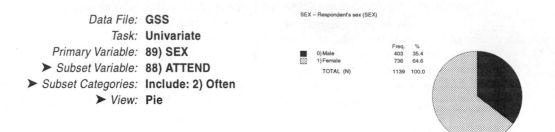

Data File:	**GSS**	
Task:	**Univariate**	
Primary Variable:	**89) SEX**	
➤ Subset Variable:	**88) ATTEND**	
➤ Subset Categories:	**Include: 2) Often**	
➤ View:	**Pie**	

The option for selecting a subset variable is located on the same screen you use to select other variables. For this example, select 88) ATTEND as a subset variable. A window will appear that shows you the categories of the subset variable. Mark the box 2) OFTEN as your subset category and then choose the "include" option. Click [OK] and continue as usual. (Note: Subset variables remain selected until you manually delete them or until you return to the main menu.)

The resulting pie chart contains information on only those who were coded 2 on 88) ATTEND—those who attend religious services often. Among those who attend church frequently, only 35.4 percent are males. So we see one way in which this sample is quite biased: Males are likely to be underrepresented. We can think of several other ways in which such a sample would probably be biased. We might expect younger individuals who are single to be underrepresented. Or perhaps individuals from different regions will be differentially represented.

Even if you selected a proper probability sample of churchgoers, you could not generalize your results to the general population. In fact, if you simply collected data at some churches or collected data on a particular Sunday, you would not have a proper sample of churchgoers and you could not even generalize your results to churchgoers.

Another form of sampling bias can result from the type of interviewing process that is selected. Some surveys, such as the U.S. General Social Survey, use face-to-face interviews, whereas other surveys use telephone interviews. Let's see if there is any serious bias generated by using telephone interviews. First, using the UNIVARIATE task, determine what percentage of the respondents have telephones.

Data File: **GSS**
Task: **Univariate**
➤ Primary Variable: **97) PHONE**
➤ View: **Pie**

If you are continuing from the previous example, make sure that you have deleted the subset variable that was used in the prior example. Again, subset variables remain selected until you manually delete them or until you return to the main menu. Remember, the [Clear All] button deletes all variable selections.

We can see that 114 cases—only 4.2 percent of the sample—did not have a phone. If NORC had conducted phone rather than personal interviews, the interviewers would not have been able to reach these individuals who didn't have phones. While this small percentage of cases would not make much difference in the results, let's nevertheless investigate who we are likely to undersample if we select only cases who have telephones. Perhaps those with low incomes would be undersampled, since they are probably somewhat less likely to have telephones. Let's first look at the distribution of income in the total sample.

Data File: **GSS**
Task: **Univariate**
➤ Primary Variable: **105) R.INCOME**
➤ View: **Pie**

As we can see, each category has about one-third of the sample. Now let's see what will happen if we look at the income distribution of only those individuals who have telephones.

Data File: **GSS**
Task: **Univariate**
Primary Variable: **105) R.INCOME**
➤ *Subset Variable:* **97) PHONE**
➤ *Subset Categories:* **Include: 1) Phone**
➤ *View:* **Pie**

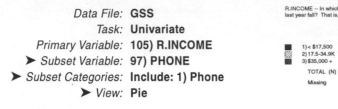
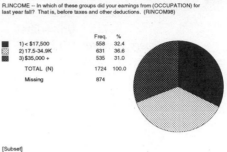

R.INCOME -- In which of these groups did your earnings from (OCCUPATION) for last year fall? That is, before taxes and other deductions. (RINCOM98)

		Freq.	%
■	1) < $17,500	558	32.4
▨	2) 17.5-34.9K	631	36.6
▦	3) $35,000 +	535	31.0
	TOTAL (N)	1724	100.0
	Missing	874	

[Subset]

Only cases with phones will be included in the distribution. We can see that excluding cases with no telephone makes only a small difference in the income distribution. For example, the percentage in the lowest income category changes from 33.6 to 32.4, only a little more than one percentage point.

Similarly, we might expect older individuals to be undersampled in a phone survey. Let's look at the distribution of age in the total sample.

Data File: **GSS**
Task: **Univariate**
➤ *Primary Variable:* **102) OVER 50?**
➤ *View:* **Pie**

OVER 50? -- IS RESPONDENT OVER 50? (AGE)

		Freq.	%
■	1) Under 50	1804	63.8
▨	2) 50 & over	1024	36.2
	TOTAL (N)	2828	100.0
	Missing	4	

Again, make sure you delete the subset variable that was selected in the previous example.

Let's now look at the distribution of age among respondents with telephones.

Data File: **GSS**
Task: **Univariate**
Primary Variable: **102) OVER 50?**
➤ *Subset Variable:* **97) PHONE**
➤ *Subset Categories:* **Include: 1) Phone**
➤ *View:* **Pie**

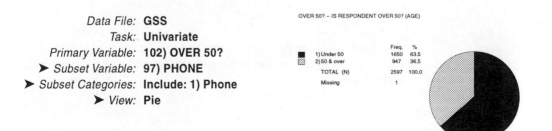

OVER 50? -- IS RESPONDENT OVER 50? (AGE)

		Freq.	%
■	1) Under 50	1650	63.5
▨	2) 50 & over	947	36.5
	TOTAL (N)	2597	100.0
	Missing	1	

[Subset]

When individuals with phones are excluded, the change is less than 1 percentage point in either of these categories. This suggests that conducting interviews over the telephone rather than using personal interviews would not have a serious effect on the results, assuming that a proper probability sample were selected (and that the nonresponse rate did not substantially increase).

Bias also can be introduced when cases selected for the sample refuse to participate. For example, in the 1998 General Social Survey, 24 percent of the individuals in the selected sample failed to complete the interview for one reason or another. To the extent that these individuals differ from those who completed the interview, the sample will be biased. Or, if a sizable number of individuals refuses to answer a particular question, any analysis involving that question may fail to reflect the population parameters. Occasionally, individuals are dropped because their answers are clearly frivolous. Sometimes, interviews are not completed because respondents become too hostile or are obviously incapable of understanding the questions. In telephone interviews, a respondent may simply hang up in the middle of the interview.

In order to help identify cases who may be providing incorrect answers, interviewers are frequently asked to provide information about the general ambience of the interview. For example, the interviewers for the GSS were asked to rate each respondent's level of comprehension. Let's look at the results.

Data File: **GSS**

Task: **Univariate**

➤ Primary Variable: **5) COMPREND**

➤ View: **Pie**

COMPREND -- Was respondent's understanding of the questions good, fair or poor? (COMPREND)

		Freq.	%
■	1) Good	2377	84.8
▨	2) Fair	360	12.8
▦	3) Poor	67	2.4
	TOTAL (N)	2804	100.0
	Missing	28	

Only a small percentage (2.4%) of respondents had difficulty understanding the questions. The likely effect of retaining individuals with poor understanding is introducing random "noise" into the survey—their answers are more likely to contain a random component than are answers of other respondents.

In addition, interviewers were asked to report whether respondents were cooperative or hostile. Let's look at the results.

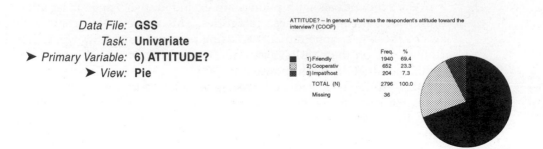

Data File: **GSS**
Task: **Univariate**
➤ Primary Variable: **6) ATTITUDE?**
➤ View: **Pie**

ATTITUDE? -- In general, what was the respondent's attitude toward the interview? (COOP)

		Freq.	%
■	1) Friendly	1940	69.4
▨	2) Cooperativ	652	23.3
▨	3) Impat/host	204	7.3
	TOTAL (N)	2796	100.0
	Missing	36	

The percentage who were impatient or hostile is also very small.

Even a properly selected sample will not be perfectly accurate in estimating population parameters. Generalizing from samples to population is a probabilistic, not a deterministic, process. For example, based on the GSS sample, our best estimate of the percent of males in the population is 43.5. Even if the response rate had been perfect, the true population parameter would likely be somewhat different from this estimate. If we select another sample of 2,832 cases using similar procedures, the percent of males is likely to be slightly different. If we select yet another sample, we will get yet another estimate. In a statistics course, you will learn how to evaluate the accuracy of such estimates.

Now let's go through the process of selecting a simple random sample using a table of random digits. While computer programs are often used to generate a random sample, the process of selecting a random sample using a table of random digits will give you a greater understanding of the basic ideas involved.

Suppose you had a list of all police officers who are members of a particular association and there were 82,000 names on this list. You want a random sample of 1,200 of these police officers for some research.

Step 1: Give each individual on the list a unique number with the same number of digits.

Here, each police officer will have a five-digit number because there are five digits in the highest number (82000). Each number will be unique—no two people will have the same number. The first person has the number 00001, the second person has the number 00002, and so on.

Step 2: Obtain a table of random digits and blindly point to a spot in the table to obtain a starting point.

A table of random digits is a table of digits set up in such a way that there is no pattern to the digits in terms of magnitude. The probability of a particular digit appearing at a particular point is the same as the probability of any other digit being there. If you blindly pick a starting point and select the first five digits from that point, the probability of any five-digit number appearing is the same as the probability of any other five-digit number being there. Thus, there is no pattern (no bias) to the order of the numbers.

Where do you get a table of random digits? Most statistics textbooks have a table of random digits in the back. There are also computer programs that can generate a list of random digits for you. The random digits table on the next page contains a series of three-digit random numbers generated by the full version of MicroCase. Note that the number 875 is emphasized. Let's suppose that this is where your finger landed when you picked the starting point.

Step 3: From the starting point, select numbers which have the same number of digits as the numbers assigned to the population.

In this situation, we need five-digit numbers because the numbers assigned to the list of police officers range from 00001 to 82000. Thus, from the starting point (875) and reading across, you will group the digits into five-digit numbers. It doesn't matter that the numbers in the columns are now set up in terms of three digits. Therefore, we have five-digit numbers as follows:

87599   19113   49583   63726   13382   32034   63626   etc.

Step 4: If a particular five-digit number falls within the range of numbers assigned to the population, the person with that number becomes part of the sample. If the number is outside the range of numbers assigned to the population, skip it and go to the next number. Continue this process until you have the number of people you need for the sample.

Here the first person in the sample is the one who has the number 87599. The second person is number 19113, the third is number 49583, and so on. We would continue this process until we had the number of people needed for the sample (although the one-page table of random digits we've used here would not be adequate). If we came up with the same number twice, we would ignore it the second time and continue.

Using this process, there is no bias in the list of people to be included in the sample. Each person on the population list has exactly the same chance (1 chance out of 82,000) of being selected for the sample.

Your turn.

RANDOM DIGITS GENERATED THROUGH MICROCASE

303	693	242	530	472	619	893	262	770	139	345	420	518
631	047	856	785	777	097	072	272	647	528	059	344	170
051	738	**875**	991	911	349	583	637	261	338	232	034	636
268	345	421	805	045	476	454	796	371	900	962	021	656
119	665	412	752	974	632	102	248	119	798	461	578	800
248	716	242	687	385	993	063	040	828	427	582	289	651
257	217	285	746	990	107	397	407	281	262	298	197	466
735	694	744	415	346	116	926	518	870	529	106	787	603
722	517	035	555	700	223	186	234	196	599	253	764	378
679	722	849	628	560	942	556	874	126	217	785	525	684
492	771	282	119	654	348	931	776	046	563	205	351	312
693	569	547	953	039	372	542	040	840	861	089	990	476
106	526	503	219	820	763	937	029	738	151	436	408	779
007	041	066	053	855	640	757	662	393	163	086	083	676
322	029	018	843	057	553	218	150	818	166	008	464	007
012	399	429	034	221	116	261	958	060	975	812	539	634
971	471	831	459	353	705	287	403	722	283	126	349	827
191	149	411	127	275	542	429	402	760	594	695	749	212
682	128	248	481	992	914	282	591	011	649	391	240	603
448	086	194	788	076	082	077	899	848	455	281	132	483
368	710	420	453	055	908	073	042	504	307	111	050	619
143	783	241	634	611	882	626	351	221	380	814	456	068
647	417	214	749	905	290	275	396	459	957	061	948	587
340	894	177	870	830	008	590	781	267	778	443	781	125
586	510	076	792	566	217	300	751	211	851	362	041	162

NAME: _____

COURSE: _____

DATE: _____

EXERCISE

4

WORKSHEET

1. The difference between population values for a variable and the sample values for the same variable is *sampling error*. For example, if the average age in the population of a state is 45.0 and the average age from a sample of this population is 46.5, then the sampling error is 1.5.

   To demonstrate this, we combined all the cases from the 1972 to 1998 General Social Surveys—a total of 38,116 people—to serve as a "population." Then we took a random sample of 800 of these respondents using random digits generated through the full version of MicroCase. We will use two variables (education and number of siblings) to demonstrate.

   Below are the averages for education and number of siblings for the population (the combined samples) and the sample. For each situation, specify the sampling error.

	Population	Sample	Sampling Error
Education	12.44	12.34	_____
Number of siblings	4.05	4.08	_____

   Would you say that the sample did a good job of representing accurately the population for these two variables? (Circle one.)   Yes   No

   When the GSS was first done in 1972, the average (mean) education level was 11.33 years. An examination of GSS results over the years since then indicates increases over time in the average number of years of school completed. In the 1996 GSS, the average was 13.36 years of school. Assuming that there was no great change in education level between 1996 and 1998, we might expect that the 1998 sample would produce results that are very similar to the 1996 results. If not, then this would suggest that the sampling procedures were not very reliable. Let's check the 1998 results.

   ➤ *Data File:* **GSS**
   ➤ *Task:* **Univariate**
   ➤ *Primary Variable:* **114) EDUCATION!**
   ➤ *View:* **Statistics (Summary)**

Remember that the mean for the 1996 GSS is: 13.36 years

Write in the mean for education in the 1998 GSS here.   _____ years

How much difference is there between the means for 1996
and 1998?   _____ years

Given the results above, what does this suggest about the reliability of the
sampling procedures used in the GSS?

2. Suppose that you were doing a survey and that, instead of selecting a
   random sample and using telephone or personal interviews, you decided to
   find your respondents at the local tavern. Using the GSS data, we can see
   how those who go to bars twice a month or more often differ from those
   who don't go to bars.

> Data File: **GSS**
> Task: **Univariate**
> ➤ Primary Variable: **9) SOCBAR**
> ➤ View: **Pie**

a. Fill in the following:

Percentage

1–/month   _____%

2+/month   _____%

b. Perhaps females would be undersampled if we selected the sample from
   bars. Let's first check the percentages of males and females and fill in the
   table that follows.

> Data File: **GSS**
> Task: **Univariate**
> ➤ Primary Variable: **89) SEX**
> ➤ View: **Pie**

Percentage

Male   _____%

Female   _____%

In the total sample, what is the percentage of females?   _____%

c. Let's compare the distribution of sex in the total sample with the subset who frequent bars. Fill in the table that follows.

> Data File: **GSS**
> Task: **Univariate**
> Primary Variable: **89) SEX**
> ➤ Subset Variable: **9) SOCBAR**
> ➤ Subset Categories: **Include: 2) 2+/month**
> ➤ View: **Pie**

Subset with value 2+/month

	Percentage
Male	_____%
Female	_____%

Among those who go to bars frequently, what
percentage is female?   _____%

What does this suggest about the distribution of sex among a sample drawn from local taverns? (Support your answer with results from the preceding analyses.)

d. People who frequent bars probably differ in many other ways from the general population. Select a variable of your choice (ideally a variable that describes a "demographic" characteristic, rather than an "attitudinal" variable) that you think will be different for people who go to bars. Then conduct a similar analysis to that done above.

Name of the variable: _____

Description of the variable:

_____

_____

e. How do you expect this variable to be different in terms of people who go to bars and the general population?

> Data File: **GSS**
> Task: **Univariate**
> ➤ Primary Variable: **(the variable you selected above)**
> ➤ View: **Pie**

Provide the distribution of the variable in the *total* sample—use only as many rows as necessary.

Category Name	Percentage
_____	_____%
_____	_____%
_____	_____%
_____	_____%
_____	_____%
_____	_____%

f. Obtain the distribution of the variable in the subset that goes to bars frequently. In addition to filling in the results below, attach a printout of your analysis. (If your computer is not connected to a printer or if you have been instructed not to use the printer, just skip these printing instructions.)

> Data File: **GSS**
> Task: **Univariate**
> Primary Variable: **(the variable you selected)**
> ➤ Subset Variable: **9) SOCBAR**
> ➤ Subset Categories: **Include: 2) 2+/month**
> ➤ View: **Pie**

Subset with value 2+/month

Category Name	Percentage
_____	_____%
_____	_____%
_____	_____%
_____	_____%
_____	_____%
_____	_____%

   g. Discuss the implications of these results for obtaining your respondents from taverns. (Support your answer with the results of your analysis.)

3. You can find many online surveys on the Internet in which people express views about political or social issues. (To see some, go to http://bsuvc.bsu.edu/~00amcorbett/online.htm). Usually, you answer the survey questions and then click a "submit" button. Then the screen presents the results of the survey based on those who have taken it so far.

Are the "samples" involved in this process representative of the general public? The NES survey includes a question which asks: "Do you have access to the Internet or the World Wide Web ('the Web')?" Look at the results for this question.

>   *Data File:* **NES**
>   *Task:* **Univariate**
> *Primary Variable:* **38) INTERNET**
>   *View:* **Pie**

   a. The NES sample is based on adults age 18 and over in the United States. What percentage of adults in the United States said that they have access to the Internet?

                                                _____%

b. Are the people who have access to the Internet similar to the people who do
   not have access? Let's compare the two groups in terms of education level.

	Data File:	**NES**
	➤ Task:	**Cross-tabulation**
	➤ Row Variable:	**38) INTERNET**
	➤ Column Variable:	**4) EDUC CATEG**
	➤ View:	**Tables**
	➤ Display:	**Column %**

	11 or less	12 years	1 – 3 colleg	4 or + col
No	_____%	_____%	_____%	_____%
Yes	_____%	_____%	_____%	_____%

Among adults in the United States, are less-educated
people or more-educated people more likely to have
Internet access? (Circle one.)   Less-educated people

                                                                 More-educated people

c. Let's examine whether race is related to Internet access.

	Data File:	**NES**
	Task:	**Cross-tabulation**
	Row Variable:	**38) INTERNET**
	➤ Column Variable:	**6) RACE**
	➤ View:	**Tables**
	➤ Display:	**Column %**

	White	Black	Nativ Amer	Asian
No	_____%	_____%	_____%	_____%
Yes	_____%	_____%	_____%	_____%

Which of these groups is least likely to have Internet
access? (Circle one.)   White Americans

                                                                 African Americans

                                                                 Native Americans

                                                                 Asian Americans

d. Are people who have Internet access different from those who do not in terms of views on social and political issues? We're now going to use Internet access as an independent variable to see whether it helps to explain variation in a social issue. To begin, read the variable description for 70) WOMEN EQL. Then do the following analysis and fill in the column percentages in the table that follows.

> Data File: **NES**
> Task: **Cross-tabulation**
> ➤ Row Variable: **70) WOMEN EQL**
> ➤ Column Variable: **38) INTERNET**
> ➤ View: **Tables**
> ➤ Display: **Column %**

	No	Yes
Woman home	_____ %	_____ %
Middle	_____ %	_____ %
Equal role	_____ %	_____ %

Which group is more traditional on this sexual equality issue? (Circle one.)

1. Those who have Internet access

2. Those who do not have Internet access

Given these results, an online Internet Survey might give the impression that the public is (Circle one.)

1. somewhat more liberal on the sexual equality issue than it actually is.

2. somewhat more conservative on the sexual equality issue than it actually is.

e. On the basis of all the preceding analysis, would it be safe to conclude that people who have Internet access are very similar to people who do not have Internet access? (Circle one.)   Yes   No

Explain your answer based on the preceding analysis.

4. Now let's use the GSS data file to investigate how respondents' education level is related to comprehension of questions.

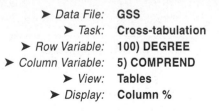

> ➤ *Data File:* **GSS**
> ➤ *Task:* **Cross-tabulation**
> ➤ *Row Variable:* **100) DEGREE**
> ➤ *Column Variable:* **5) COMPREND**
> ➤ *View:* **Tables**
> ➤ *Display:* **Column %**

a. Fill in the column percentages in the following table.

	Good	Fair	Poor
Not hi school	_____%	_____%	_____%
High sch	_____%	_____%	_____%
Some coll	_____%	_____%	_____%

b. Assuming that individuals who fail to understand the questions are most likely to break off an interview, how would the distribution of educational attainment in the sample be affected?

5. On presidential election day, newscasters report results of exit polls based on surveys of voters leaving the polling area. In addition to reporting results on votes, the newscasters might also report the views of voters on various political issues and they might discuss the demographic characteristics of voters. These exit poll results are appropriate for describing the relevant population (voters), but the results might not be as accurate in reflecting the views of the general public—both voters and nonvoters. Nonvoters might differ from voters in terms of political attitudes and demographic characteristics.

Let's use the NES data file in order to investigate this. (Note: Many who claimed that they voted did not actually vote. Although this does cause a problem for the analysis we're doing here, we will ignore this issue for the present.)

a. Let's first obtain the distribution of attitudes for 74) JOB GUAR.

> ➤ *Data File:* **NES**
> ➤ *Task:* **Univariate**
> ➤ *Primary Variable:* **74) JOB GUAR**
> ➤ *View:* **Pie**

Fill in the following information.

	Percentage
1) On own	_____%
2) Middle	_____%
3) Guarantee	_____%

b. Now let's look at the distribution of attitudes on this question for just those who said they voted.

> *Data File:* **NES**
> *Task:* **Univariate**
> *Primary Variable:* **74) JOB GUAR**
> ➤ *Subset Variable:* **41) VOTE96**
> ➤ *Subset Categories:* **Include: 1) Yes**
> ➤ *View:* **Pie**

Fill in the information in the following table.

Subset with value YES

	Percentage
1) On own	_____%
2) Middle	_____%
3) Guarantee	_____%

c. Compare the distribution for the total sample with the distribution for those who said they voted. If we used a sample of voters to represent the views of the general public, what effect would it have on our impression of public attitudes concerning this issue? First, would we overestimate or underestimate the level of support for job guarantees? Second, do you think this difference is *substantively* significant? That is, do you think the percentage difference between those who voted and those who did not vote is large enough to be concerned with? (Support your answer using the results of your analysis.)

d. Would using the views of voters to represent public opinion cause a problem in examining views on the issue of prayer in public schools? First, obtain the distribution for the total sample.

> | | |
> |---------------------:|:---------------|
> | Data File: | **NES** |
> | Task: | **Univariate** |
> | ➤ Primary Variable: | **11) SCHL PRAYR** |
> | ➤ View: | **Pie** |

Fill in the information in the following table.

	Percentage
Not allowed	_____%
Silent prayer	_____%
General prayer	_____%
Christian prayer	_____

e. Now let's look at the distribution of attitudes on this question for just those who said they voted.

> | | |
> |-----------------------:|:-----------------|
> | Data File: | **NES** |
> | Task: | **Univariate** |
> | Primary Variable: | **11) SCHL PRAYR** |
> | ➤ Subset Variable: | **41) VOTE96** |
> | ➤ Subset Categories: | **Include: 1) Yes** |
> | ➤ View: | **Pie** |

Fill in the following information:

Subset with value YES

	Percentage
Not allowed	_____%
Silent prayer	_____%
General prayer	_____%
Christian prayer	_____%

f. Compare the distribution for the total sample with the distribution for those who said they voted. If we used a sample of voters to represent the views of the general public, what effect would it have on our impression of public attitudes concerning this issue? Would we overestimate or underestimate the level of support for prayer in public schools? (Use the results of the analysis to support your answer.)

6. A researcher wants to generalize from his survey results to the population that received his questionnaire. When asked if the returned questionnaires could be considered a random sample of the population, he replies, "We sent out 100,000 questionnaires and got only 5,000 back. What could be more random than that?" Is he right? Why or why not?

5a

Control Variables

OVERVIEW

In Exercise 5a, you will learn how to examine a relationship between two variables while controlling for the effects of another variable—and you will see why this is important when we try to disentangle relationships among variables. Exercise 5b is an optional assignment (ask your instructor) that looks at more advanced issues on this topic. In this exercise, you will gain experience in using regression to examine the simultaneous effects of two or more independent variables on a dependent variable, and you will see how this is useful in assessing models of what causes the variance in the dependent variable.

BEFORE YOU BEGIN

Please make sure you have read Chapter 5 in the textbook and can answer the following review questions (you need not write any answers):

1. Describe the three criteria of causation (time order, etc.).

2. How are independent, dependent, intervening, and antecedent variables related to each other?

3. What is multiple causation?

4. What are causal models, and how is regression used in assessing causal models?

5. If the beta coefficient for an independent variable in regression is .50, what does this mean? (Assume that it is statistically significant.)

6. If a squared multiple regression correlation (R^2) is .80, what does this mean?

7. How can regression be useful in identifying spurious relationships?

8. How can regression help in sorting out the effects of suppressor variables?

9. What are dummy variables, and how are they used in regression?

Before we look at control variables in cross-tabulation, let's take a closer look at the variables we'll be using.

> ➤ *Data File:* **GSS**
> ➤ *Task:* **Cross-tabulation**
> ➤ *Row Variable:* **116) R.INCOME!**
> ➤ *Column Variable:* **105) R.INCOME**
> ➤ *View:* **Tables**
> ➤ *Display:* **Frequency**

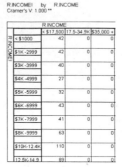

R.INCOME!   by   R.INCOME
Cramer's V: 1.000 **

		R.INCOME		
		< $17,500	17.5-34.9K	$35,000 +
R.INCOME!	< $1000	42	0	0
	$1K -2999	42	0	0
	$3K -3999	40	0	0
	$4K -4999	27	0	0
	$5K -5999	32	0	0
	$6K -6999	43	0	0
	$7K -7999	41	0	0
	$8K -9999	63	0	0
	$10K-12.4K	110	0	0
	12.5K-14.9	89	0	0

Although you are generally instructed to display column percentaging in your cross-tabulation analysis, note that here we are looking at the frequency distribution.

Variable 105) R.INCOME is based on exactly the same question as variable 116) R. INCOME!. The only difference is that variable 116 includes all answer categories made available to respondents for reporting their own annual income, whereas variable 105 was created by combining a number of these categories. The table shows you which categories of 116) R.INCOME! have been combined to form the categories for 105) R.INCOME. For example, categories from <$1000 to $15K–17.4K on variable 116) R.INCOME! were placed in the category <$17,500 on variable 105) R.INCOME.

The procedure of combining response categories for a variable is called *collapsing*—105) R.INCOME is a *collapsed* version of 116) R.INCOME!. In this data set, all variables with an exclamation point are uncollapsed versions of variables (many of these also are included in a collapsed form). The uncollapsed versions are located toward the end of the variable list. It may have crossed your mind that collapsing seems to go against the idea of trying to have as much variation as possible in our variables since variable 105 has only three categories while variable 116 has 23. Why do we do this? Obtain the following cross-tabulation table and look at the column percentages.

> *Data File:* **GSS**
> *Task:* **Cross-tabulation**
> *Row Variable:* **116) R.INCOME!**
> ➤ *Column Variable:* **114) EDUCATION!**
> ➤ *View:* **Tables**
> ➤ *Display:* **Column %**

R.INCOME!   by   EDUCATION!
Cramer's V: 0.143 **
Warning: Potential significance problem. Check row and column totals.

		EDUCATION!					
		No school	2nd grade	3rd grade	4th grade	5th grade	6th grade
R.INCOME!	< $1000	0	1	0	0	0	0
		0.0%	50.0%	0.0%	0.0%	0.0%	0.0%
	$1K -2999	0	0	0	0	0	0
		0.0%	0.0%	0.0%	0.0%	0.0%	0.0%
	$3K -3999	0	0	1	0	0	1
		0.0%	0.0%	33.3%	0.0%	0.0%	25.0%
	$4K -4999	0	0	0	0	0	0
		0.0%	0.0%	0.0%	0.0%	0.0%	0.0%
	$5K -5999	0	0	0	0	0	0
		0.0%	0.0%	0.0%	0.0%	0.0%	0.0%
	$6K -6999	0	0	0	0	0	0
		0.0%	0.0%	0.0%	0.0%	0.0%	0.0%
	$7K -7999	0	0	1	0	0	0
		0.0%	0.0%	33.3%	0.0%	0.0%	0.0%
	$8K -9999	0	0	0	0	0	0
		0.0%	0.0%	0.0%	0.0%	0.0%	0.0%

You need to use the scroll bar to move around in this huge table. Even a very experienced professional analyst would not be able to make much sense of this table. Many of the columns have very few cases, so the percentages for the most part are meaningless. This is why we collapse variables when using tabular analysis—to get enough cases in the categories to be able to interpret the results.

Because we will be using cross-tabulation in this exercise, virtually all of the variables are in collapsed form. While collapsing will change the *distribution* of the variable, such changes rarely affect *relationships* with other variables as we saw in Exercise 2b. (In Chapter 7 of the text, you'll learn the principles of collapsing variables.)

Now let's look at an example of a spurious relationship. We will begin by cross-tabulating sexual frequency by marital status and filling in the column percentages in the table that follows.

Data File: **GSS**
Task: **Cross-tabulation**
➤ *Row Variable:* **11) SEX FREQ**
➤ *Column Variable:* **96) MARITAL**
➤ *View:* **Tables**
➤ *Display:* **Column %**

SEX FREQ   by   MARITAL
Cramer's V: 0.389 **

		Married	Div/widowd	Nev.marry	Missing	TOTAL
SEX FREQ	-Monthly	141	358	192	0	691
		12.8%	54.7%	34.0%		29.8%
	Monthly+	961	296	372	0	1629
		87.2%	45.3%	66.0%		70.2%
	Missing	244	168	99	1	512
	TOTAL	1102	654	564	1	2320
		100.0%	100.0%	100.0%		

The row variable tells us how often a respondent has had sex during the previous 12 months. In this form, the variable has been collapsed into those who had sex less often than once a month and those who had sex at least once a month or more often. It should surprise no one that married people were more apt to have sex at least once a month than were people in the other two categories. However, it does seem unusual that the difference between the never married and the widowed and divorced group should be so great (around a 21 percent difference). Let's focus on just those two groups.

Data File: **GSS**
Task: **Cross-tabulation**
Row Variable: **11) SEX FREQ**
Column Variable: **96) MARITAL**
➤ *Subset Variable:* **96) MARITAL**
➤ *Subset Categories:* **Exclude: 1) Married**
➤ *View:* **Tables**
➤ *Display:* **Column %**

SEX FREQ   by   MARITAL
Cramer's V: 0.207 **

		Div/widowd	Nev.marry	TOTAL
SEX FREQ	-Monthly	358	192	550
		54.7%	34.0%	45.2%
	Monthly+	296	372	668
		45.3%	66.0%	54.8%
	Missing	168	99	267
	TOTAL	654	564	1218
		100.0%	100.0%	

The option for selecting a subset variable is located on the same screen you use to select other variables. For this example, select 96) MARITAL as a subset variable. A window will

appear that shows you the categories of the subset variable. Mark the box 1) MARRIED and then choose the [Exclude] option. (This means that all other categories for this variable will be included in the analysis.) Click [OK] and continue on as usual. Remember, subset variables remain selected until you manually delete them or until you return to the main menu.

We can now focus on the difference between the divorced/widowed and never-married groups because our analysis excludes people who are married. This table shows a percentage difference of 20.7 in the first row, and the N's (number of cases) reported at the bottom are large enough that we can have some confidence in the stability of the percentages. You can see that Cramer's V is .207 and is highly significant.

We have now observed a relationship between marital status and frequency of sex among those who are not married. But could this relationship be spurious? Is there some other variable that could be creating this difference? If so, this other variable must precede these two variables in time. Perhaps age might be this variable. Age has an effect on both of these variables—the never-married group is probably younger and younger individuals are probably more likely to have sex.

Let's diagram this argument:

AMONG THE NOT MARRIED

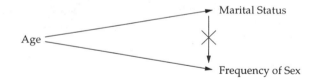

The position of the variables from left to right indicates time order. The arrows indicate causal relationships. The arrow at the right with the X across it represents a relationship that disappears when we control for age. Now let's test this argument by controlling for age.

Data File:	**GSS**
Task:	**Cross-tabulation**
Row Variable:	**11) SEX FREQ**
Column Variable:	**96) MARITAL**
Subset Variable:	**96) MARITAL**
Subset Categories:	**Exclude: 1) Married**
➤ *Control Variable:*	**102) OVER 50?**
➤ *View:*	**Tables (Under 50)**
➤ *Display:*	**Column %**

SEX FREQ   by   MARITAL

Controls: OVER 50?: Under 50
Cramer's V: 0.011

	MARITAL		
		Div/widowd	Nev marry	TOTAL
SEX FREQ	-Monthly	93	159	252
		29.7%	30.8%	30.4%
	Monthly+	220	358	578
		70.3%	69.2%	69.6%
	Missing	51	87	138
	TOTAL	313	517	830
		100.0%	100.0%	

If you are continuing from the previous example, and you haven't returned to the main menu, simply add a control variable to your previous variable selections. The option for selecting a control variable is located on the same screen you use to select other variables. For this example, select 102) OVER 50? as a control variable and then click [OK] to continue. Separate tables for each of the OVER 50? categories will now be shown for the SEX FREQ and MARITAL cross-tabulation.

The first table will show the relationship between marital status and frequency of sex for those unmarried respondents under 50. As we can see, there is virtually no difference between the divorced/widowed and never-married groups among those under 50. You can see that Cramer's V is .011 and clearly not significant.

Let's move on to the second control table, which limits the analysis to those who are 50 and over.

Data File:	**GSS**
Task:	**Cross-tabulation**
Row Variable:	**11) SEX FREQ**
Column Variable:	**96) MARITAL**
Subset Variable:	**96) MARITAL**
Subset Categories:	**Exclude: 1) Married**
Control Variable:	**102) OVER 50?**
➤ View:	**Tables (50 & Over)**
➤ Display:	**Column %**

SEX FREQ   by   MARITAL

Controls: OVER 50?: 50 & over
Cramer's V: 0.060

		MARITAL		
		Div/widowd	Nev.marry	TOTAL
SEX FREQ	-Monthly	265	33	298
		77.9%	70.2%	77.0%
	Monthly+	75	14	89
		22.1%	29.8%	23.0%
	Missing	116	11	127
	TOTAL	340	47	387
		100.0%	100.0%	

If you are continuing from the previous example, click the appropriate button at the bottom of the task bar to look at the second (or "next") partial table for those who are 50 and over.

As can be seen from this table, among those 50 and over, there is little difference in sexual frequency between those who are divorced/widowed and those who have never married. You can see that Cramer's V is .060 and not significant. (While this relationship might be statistically significant if we had a much larger sample, the relationship would still be too weak to be of any importance.) Thus, the relationship between marital status (excluding married individuals) and frequency of sex would appear to be spurious, produced by variations in age. Widowed and divorced persons do not have sex less frequently than do never-married people *in their age group*! The initial differences between the groups are the result of the fact that, as a group, people who are widowed or divorced are far more likely to be over 50 than are persons who have never married, most of whom are young adults who soon will marry. So our diagram is an accurate summary of the relationships.

These results are unusually clean. Typically, when control variables are used, there will be some "bounce," or random fluctuation, in the subtables because of the small numbers of cases in some columns. We can see more of the messiness of real analysis in other examples.

Abortion has become one of the most controversial issues in American society, and the political parties have in recent times taken different positions on the issue: The Republican Party has become associated with a more restrictive view on abortion, and the Democratic Party has become associated with a less restrictive view on abortion. Is this political party difference on the abortion issue represented in the general population? Let's find out, using variable 52) ABORT POOR (allow an abortion if the family has a very low income and cannot afford any more children). Note: Before you start the instructions in the following software guide, it would be a good idea to click on the [Clear All] button to clear all variables from the previous analysis.

Data File:	**GSS**
Task:	**Cross-tabulation**
➤ *Row Variable:*	**52) ABORT POOR**
➤ *Column Variable:*	**22) POLPARTY**
➤ *View:*	**Tables**
➤ *Display:*	**Column %**

ABORT POOR   by   POLPARTY
Cramer's V: 0.129 **

ABORT POOR		POLPARTY				
		Democrat	Independ	Republican	Missing	TOTAL
	No	329	339	305	21	973
		52.0%	52.6%	66.7%		56.1%
	Yes	304	306	152	27	762
		48.0%	47.4%	33.3%		43.9%
	Missing	334	425	266	24	1049
	TOTAL	633	645	457	72	1735
		100.0%	100.0%	100.0%		

These results show that there is a political party difference in the general public on the issue of abortion: Republicans are more restrictive of abortion in this particular situation than Democrats and Independents are. Cramer's V is .129 and the relationship is statistically significant. Why does this relationship exist? What variable might intervene between political party preference and abortion views that would bring about this relationship? Let's use political ideology as an intervening variable. While the political parties might attract different kinds of people in terms of political ideologies, the political parties might also shape the political ideologies of their members. Political party identification is to a great extent learned during early childhood, and it might shape the person's ideological framework, and that ideological framework could in turn shape the person's view on abortion. However, if this process did not occur for certain people, then we would not expect that party identification and abortion views would be related among these people. Thus, we can speculate that if we control for the intervening variable, political ideology, there would be little or no relationship left between political party preference and abortion views. Putting this differently, political party preference would affect views on abortion only if political party preference were filtered through political ideology identifications. We can diagram this argument as follows:

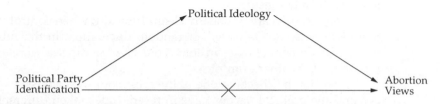

Notice that, when we are dealing with intervening variables, the variable comes between the other two variables in time. Again the arrows represent causal relationships. We have argued that political ideology is an intervening variable between political party identification and abortion views. So, in this situation, we expect the relationship between political party identification and abortion views to disappear when we control for political ideology. Let's check this idea using 23) POLVIEW as an indicator of political ideology.

Data File:	**GSS**
Task:	**Cross-tabulation**
Row Variable:	**52) ABORT POOR**
Column Variable:	**22) POLPARTY**
➤ *Control Variable:*	**23) POLVIEW**
➤ *View:*	**Tables (Liberal)**
➤ *Display:*	**Column %**

ABORT POOR by POLPARTY
Controls: POLVIEW: Liberal
Cramer's V: 0.071

		Democrat	Independ	Republican	Missing	TOTAL
ABORT POOR	No	99	69	25	4	193
		37.6%	37.5%	49.0%		38.8%
	Yes	164	115	26	12	305
		62.4%	62.5%	51.0%		61.2%
	Missing	127	96	31	4	258
	TOTAL	263	184	51	20	498
		100.0%	100.0%	100.0%		

These results show that among liberals, the linkage between political party and abortion views is weaker (Cramer's V = .071 here but it was .129 for the total sample) than it was before and the relationship here is not statistically significant. Now let's examine the relationship among moderates.

Data File:	**GSS**
Task:	**Cross-tabulation**
Row Variable:	**52) ABORT POOR**
Column Variable:	**22) POLPARTY**
Control Variable:	**23) POLVIEW**
➤ *View:*	**Tables (Moderate)**
➤ *Display:*	**Column %**

ABORT POOR by POLPARTY
Controls: POLVIEW: Moderate
Cramer's V: 0.020

		Democrat	Independ	Republican	Missing	TOTAL
ABORT POOR	No	129	139	66	8	334
		57.8%	55.8%	57.9%		57.0%
	Yes	94	110	48	4	252
		42.2%	44.2%	42.1%		43.0%
	Missing	128	170	76	14	388
	TOTAL	223	249	114	26	586
		100.0%	100.0%	100.0%		

If you are continuing from the previous analysis, simply click the appropriate button at the bottom of the task bar to look at the second partial table for POLVIEW (i.e., those who classified themselves as moderates).

Among moderates, there is virtually no difference between Democrats, Independents, and Republicans on this abortion issue. Thus, among moderates, political party preference is not related to views on this particular abortion issue. Cramer's V is only .020 and is not statistically significant. Now let's check the relationship among conservatives.

Data File:	**GSS**
Task:	**Cross-tabulation**
Row Variable:	**52) ABORT POOR**
Column Variable:	**22) POLPARTY**
Control Variable:	**23) POLVIEW**
➤ *View:*	**Tables (Conservative)**
➤ *Display:*	**Column %**

ABORT POOR by POLPARTY
Controls: POLVIEW: Conserv.
Cramer's V: 0.078

		Democrat	Independ	Republican	Missing	TOTAL
ABORT POOR	No	82	105	206	7	393
		67.2%	66.9%	74.1%		70.6%
	Yes	40	52	72	10	164
		32.8%	33.1%	25.9%		29.4%
	Missing	69	130	154	6	359
	TOTAL	122	157	278	23	557
		100.0%	100.0%	100.0%		

If you are continuing from the previous analysis, simply click the appropriate button at the bottom of the task bar to look at the third partial table for POLVIEW (i.e., those who classified themselves as conservatives).

Here we see that, among conservatives, the differences in views on this particular abortion issue are not that great among Democrats, Independents, and Republicans. While there is still a small difference between Republicans and the other two groups, the difference is not substantial and it is not statistically significant. Thus, overall, when we control for political ideology, there is no longer a substantial relationship between political party preference and views on this abortion issue.

This example also demonstrated the "messiness" that can exist in real analysis (as opposed to neat, clear-cut hypothetical examples in which the patterns are nearly perfect). In this situation, among the liberals and the conservatives, the Republicans were still slightly different from others in their views on abortion—even though the differences were small and not statistically significant.

Let's pause for a moment to summarize the difference between spurious and intervening relationships. To test for spuriousness, we control for a variable and examine the relationship between two variables. To test for an intervening variable, we do exactly the same thing. The only difference is the time order of the variables. With a spurious relationship, the control variable precedes the other two variables. With an intervening relationship, the control variable comes between the other two variables.

These concepts are slightly complex, so let's make sure they are understood before we go on. In the first example, there was a correlation between age and marital status, as well as a correlation between age and frequency of sex. But a third relationship between marital status and frequency of sex disappeared when you controlled for age. As the model showed, age preceded both marital status and the frequency of sex, and the apparent relationship between marital status and frequency of sex was spurious.

In the second example, there appeared to be a relationship between political party identification and views on abortion (in the situation in which the family is poor and cannot afford any more children). But one reason that this link could occur is that political party identification can affect the ideological identifications of people, and ideological identifications can in turn affect their views on abortion. Thus, ideological identification was a link—an intervening variable—to the relationship between political party identification and views on abortion.

Let's look at another example of a three-variable relationship. We might reason that individuals with low family incomes are more likely to live in neighborhoods with high crime rates and, consequently, might be more worried about walking alone at night. Let's check this.

Data File: **GSS**
Task: **Cross-tabulation**
➤ Row Variable: **4) FEAR WALK**
➤ Column Variable: **104) INCOME**
➤ View: **Tables**
➤ Display: **Column %**

FEAR WALK   by   INCOME
Cramer's V: 0.161 **

		INCOME				
		< $22,500	22.5-49.9K	$50,000 +	Missing	TOTAL
FEAR WALK	Yes	268	242	154	108	664
		51.1%	39.0%	31.3%		40.6%
	No	256	379	338	106	973
		48.9%	61.0%	68.7%		59.4%
	Missing	271	319	276	115	981
	TOTAL	524	621	492	329	1637
		100.0%	100.0%	100.0%		

In fact, this appears to be the case. We can see that the higher the income category is, the less likely respondents are to say that they fear walking alone in their neighborhood at night.

But we may also think that crime is generally associated with cities. We might further think that people with low incomes are more likely to live in urban neighborhoods than are people with higher incomes. Hence, we might hypothesize that if we control for the urbanity of a neighborhood, the relationship between income and fear of walking alone in one's neighborhood at night will disappear.

In the following diagram, we have predicted that the relationship between income and fear of walking will disappear when we control for urbanity of neighborhood.

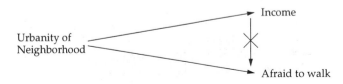

To test this idea, we would control for area of residence, 110) PLACE SIZE, and reexamine this relationship. So, repeat the previous analysis using PLACE SIZE as the control variable. You should be able to obtain each of the following tables (remember to use column percentaging for each control table).

Data File: **GSS**
Task: **Cross-tabulation**
Row Variable: **4) FEAR WALK**
Column Variable: **104) INCOME**
➤ Control Variable: **110) PLACE SIZE**
➤ View: **Tables**
➤ Display: **Column %**

FEAR WALK   by   INCOME
Controls: PLACE SIZE: Town/farm
Cramer's V: 0.052

		INCOME				
		< $22,500	22.5-49.9K	$50,000 +	Missing	TOTAL
FEAR WALK	Yes	26	26	17	12	69
		30.2%	27.4%	24.3%		27.5%
	No	60	69	53	23	182
		69.8%	72.6%	75.7%		72.5%
	Missing	52	54	42	18	166
	TOTAL	86	95	70	53	251
		100.0%	100.0%	100.0%		

FEAR WALK   by   INCOME
Controls: PLACE SIZE: Small city
Cramer's V: 0.217 *

		INCOME				
		< $22,500	22.5-49.9K	$50,000 +	Missing	TOTAL
FEAR WALK	Yes	22	24	6	10	52
		47.8%	42.1%	20.0%		39.1%
	No	24	33	24	8	81
		52.2%	57.9%	80.0%		60.9%
	Missing	26	29	13	5	73
	TOTAL	46	57	30	23	133
		100.0%	100.0%	100.0%		

FEAR WALK   by   INCOME
Controls: PLACE SIZE: City/subur
Cramer's V: 0.185 **

		INCOME				
		< $22,500	22.5-49.9K	$50,000 +	Missing	TOTAL
FEAR WALK	Yes	220	192	131	86	543
		56.1%	40.9%	33.4%		43.3%
	No	172	277	261	75	710
		43.9%	59.1%	66.6%		56.7%
	Missing	193	236	221	92	742
	TOTAL	392	469	392	253	1253
		100.0%	100.0%	100.0%		

Study each of these tables. We can see that area of residence does have a powerful effect. Consider the percentages of those in the lowest income bracket who said Yes to the FEAR WALK question: 56.1% of those in cities (or suburbs of large cities), 47.8% of those in small cities, but only 30.2% of those in small towns or farm areas. Thus, size of area of residence has a definite impact on responses to the FEAR WALK question. However, the effect of income also still exists. In all three size of place areas (town/farm, small city, city/suburb), those with higher incomes were less likely to say that there was anyplace around them where they feared to walk at night—although this relationship is not statistically significant for the town/farm category. So, in this example, both variables have an effect on the dependent variable and our initial diagram was not correct.

We could rediagram this result to reflect the observed relationships:

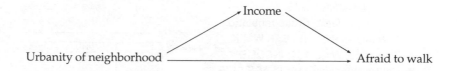

In this example, both independent variables have an effect on the dependent variable.

Your turn.

1. Let's look at the relationship between gun ownership and attendance at religious services. Be sure to read the variable descriptions before conducting your analysis.

> ➤ Data File: **GSS**
> ➤ Task: **Cross-tabulation**
> ➤ Row Variable: **18) YOUR GUN?**
> ➤ Column Variable: **88) ATTEND**
> ➤ View: **Tables**
> ➤ Display: **Column %**

a. Fill in the column percentages in the table that follows, and then obtain Cramer's V and the significance level for this relationship.

	Not often	Often
Yes	_____%	_____%
No	_____%	_____%

V = _____

Prob. = _____

Is this relationship statistically significant? (Circle one.)   Yes   No

How would you interpret these results? That is, what kind of pattern is there in the results?

b. We might argue that this is a spurious relationship and that both of these variables are affected by gender: Women are more likely to go to church, and women are less likely to own guns. Draw the diagram that would represent this argument.

c. Conduct the same analysis again but use 89) SEX as the control variable this time. Fill in the column percentages, Cramer's V, and the significance levels.

> | | |
> |---:|:---|
> | Data File | **GSS** |
> | Task: | **Cross-tabulation** |
> | Row Variable: | **18) YOUR GUN?** |
> | Column Variable: | **88) ATTEND** |
> | ➤ Control Variable: | **89) SEX** |
> | ➤ View: | **Tables** |
> | ➤ Display: | **Column %** |

Control Category: MALE

	Not often	Often
Yes	_____%	_____%
No	_____%	_____%

V = _____

Prob. = _____

Is this relationship statistically significant? (Circle one.)   Yes   No

Control Category: FEMALE

	Not often	Often
Yes	_____%	_____%
No	_____%	_____%

V = _____

Prob. = _____

Is this relationship statistically significant? (Circle one.)   Yes   No

d. Does the analysis support the diagram you drew in
   Question 1b? (Circle one.)   Yes   No

e. In your own words, provide an interpretation of these results. Cite evidence from the tables to support your answer.

2. Some religious orientations that people hold are related to their views on
   other social matters. Let's look at the relationship between biblical literalism
   and belief in the housewife role for women.

Data File	**GSS**
Task:	**Cross-tabulation**
➤ Row Variable:	**60) HOUSEWIFE**
➤ Column Variable:	**78) BIBLE1**
➤ View:	**Tables**
➤ Display:	**Column %**

a. Fill in the column percentages in the following table and obtain Cramer's
   V and the significance level.

	Actual	Inspired	Ancient Book
Agree	_____%	_____%	_____%
Disagree	_____%	_____%	_____%

V = _____

Prob. = _____

Is this relationship statistically significant? (Circle one.)   Yes   No

b. Why is it that those who interpret the Bible literally are more likely to support a housewife role for women? One explanation is that the substance of the Bible would lead to a certain kind of role for women. However, another explanation is that this relationship is spurious—that both biblical literalism and support for the traditional housewife role for women are both influenced by education levels. The suggestion is that people with lower education are more likely to interpret the Bible literally and that they are also more likely to support the traditional housewife role for women. Thus, it might be that there is no relationship between biblical literalism and support for a housewife role for women when we control for education level. Diagram this argument below.

c. Test this argument that the relationship between biblical literalism and support for a housewife role for women is spurious because both variables are affected by education level. Do the following analyses and fill in the column percentages, V, and significance level for each analysis.

> | | |
> | ---: | :--- |
> | Data File | **GSS** |
> | Task: | **Cross-tabulation** |
> | Row Variable: | **60) HOUSEWIFE** |
> | Column Variable: | **78) BIBLE1** |
> | ➤ Control Variable: | **100) DEGREE** |
> | ➤ View: | **Tables** |
> | ➤ Display: | **Column %** |

Control Category: NOT HI SCH

	Actual	Inspired	Ancient Book
Agree	_____%	_____%	_____%
Disagree	_____%	_____%	_____%

V = _____

Prob. = _____

Is this relationship statistically significant? (Circle one.)　　　Yes　　No

Control Category: HI SCH

	Actual	Inspired	Ancient Book
Agree	_____%	_____%	_____%
Disagree	_____%	_____%	_____%

V = _____

Prob. = _____

Is this relationship statistically significant? (Circle one.)　　　Yes　　No

Control Category: SOME COLL.

	Actual	Inspired	Ancient Book
Agree	_____%	_____%	_____%
Disagree	_____%	_____%	_____%

V = _____

Prob. = _____

Is this relationship statistically significant? (Circle one.)　　　Yes　　No

d. Do the results in these tables support the proposition
that the relationship between biblical literalism and support
for a housewife role for women is a spurious relationship?
(Circle one.)　　　Yes　　No

Explain your answer citing the relevant statistics.

3. Now let's use the NES data file to test some ideas.

   a. First, who would you expect to have higher personal trust of other people: those with lower incomes or those with higher incomes? (Circle one.)

People with lower incomes

People with higher incomes

   b. You can test your idea by using variables 107) TRUST? and 10) %50%50%. Make sure to read the variable descriptions before doing the analysis in the following software guide. Fill in the column percentages in the table that follows and then obtain Cramer's V and the significance level.

> ➤ *Data File:* **NES**
> ➤ *Task:* **Cross-tabulation**
> ➤ *Row Variable:* **107) TRUST?**
> ➤ *Column Variable:* **10) %50%50%**
> ➤ *View:* **Tables**
> ➤ *Display:* **Column %**

	Lower 50%	Upper 50%
Be careful	_____%	_____%
Can trust	_____%	_____%

V = _____

Prob. = _____

Is this relationship statistically significant? (Circle one.)   Yes   No

Do these results support the hypothesis you specified in Question 3a? (Circle one).   Yes   No

   c. It is possible that the relationship between income level and personal trust is spurious. One possibility is that education level affects both income level and personal trust. Draw the diagram that would represent this argument.

d. Use variable 112) COLLEGE? as a control variable to test the idea that the relationship between income and personal trust is spurious.

Data File:	**NES**
Task:	**Cross-tabulation**
Row Variable:	**107) TRUST?**
Column Variable:	**10) %50%50%**
➤ Control Variable:	**112) COLLEGE?**
➤ View:	**Tables**
➤ Display:	**Column %**

Control Category: NO COLLEGE

	Lower 50%	Upper 50%
Be careful	_____ %	_____ %
Can trust	_____ %	_____ %

V = _____

Prob. = _____

Is this relationship statistically significant? (Circle one.)   Yes   No

Control Category: YES

	Lower 50%	Upper 50%
Be careful	_____ %	_____ %
Can trust	_____ %	_____ %

V = _____

Prob. = _____

Is this relationship statistically significant? (Circle one.)   Yes   No

e. Do these results support the argument that the relationship between income and personal trust is spurious, that the link exists only because both income and personal trust are related to education level? (Circle one.)   Yes   No

Explain your answer, citing the appropriate results to support your conclusion.

f. Based on the preceding results, draw the diagram that shows the relationships of income level and education level to the dependent variable personal trust.

4. a. Let's examine the following proposition: People with a college degree are more likely to participate in politics than people who do not have a college degree. We will use variable 55) TALK POLTC as an indicator of political participation. Do the following analysis and fill in the column percentages, Cramer's V, and the significance level.

Data File:	**NES**
Task:	**Cross-tabulation**
➤ Row Variable:	**55) TALK POLTC**
➤ Column Variable:	**112) COLLEGE?**
➤ View:	**Tables**
➤ Display:	**Column %**

	No college	Yes
No	_____%	_____%
Yes	_____%	_____%

$$V = \underline{\hspace{2cm}}$$

$$\text{Prob.} = \underline{\hspace{2cm}}$$

Is this relationship statistically significant? (Circle one.)   Yes   No

b. We see that having a college degree is associated with higher political participation in this situation. However, the question arises as to why this is so. We can speculate that having a college degree leads to increased political knowledge which, in turn, leads to increased political participation. Draw the diagram for this argument below.

c. Now test the argument, using variable 80) POL INFO as a control variable. Fill in the column percentages, Cramer's V, and significance level for each analysis.

Data File:	**NES**
Task:	**Cross-tabulation**
Row Variable:	**55) TALK POLTC**
Column Variable:	**112) COLLEGE?**
➤ *Control Variable:*	**80) POL INFO**
➤ *View:*	**Tables**
➤ *Display:*	**Column %**

Control Category: LOW

	No college	Yes
No	_____%	_____%
Yes	_____%	_____%

$$V = \underline{\hspace{2cm}}$$

$$\text{Prob.} = \underline{\hspace{2cm}}$$

Is this relationship statistically significant? (Circle one.)   Yes   No

Control Category: AVERAGE

	No college	Yes
No	_____%	_____%
Yes	_____%	_____%

V = _____

Prob. = _____

Is this relationship statistically significant? (Circle one.)   Yes   No

Control Category: HIGH

	No college	Yes
No	_____%	_____%
Yes	_____%	_____%

V = _____

Prob. = _____

Is this relationship statistically significant? (Circle one.)   Yes   No

d. Do the results support the proposition that political knowledge
is an intervening variable in the relationship between having a
college education and political participation? (Circle one.)   Yes   No

Explain your answer, citing the relevant results.

EXERCISE 5b

Causal Models

In the previous exercise, we saw how quickly cases can evaporate in cross-tabular analysis, making interpretations extremely difficult. This happens even when we work with collapsed variables. In this exercise, we'll look at another technique, regression analysis, that can be used for the same type of analysis. This technique uses statistical procedures to unravel the influence of different independent variables and does not have the problem of running out of cases.

As you select the variables in the following guide, look at the variable descriptions. The exclamation points in the names indicate that these are the uncollapsed forms of the variables and therefore these variables preserve the maximum amount of variation. In regression analysis, the more variation the better. Notice that for the GSS variable SEX FREQ! (sexual frequency during the last twelve months) there are seven categories, ranging from "not at all" to "more than three times a week." However, this is an ordinal variable rather than being interval or ratio. Ideally, we use only interval or ratio variables with regression, but we'll assume that these variables are close enough to interval variables to be acceptable.

> *Data File:* **GSS**
> *Task:* **Regression**
> *Dependent Variable:* **122) SEX FREQ!**
> *Independent Variable:* **113) AGE!**
> *View:* **Graph**

Multiple R-Squared = 0.193**

AGE! ——————— BETA = -0.440** ——————— SEX FREQ!
(r = -0.440)

We can see that age has a substantial effect on sexual frequency. The older people are, the lower their sexual frequency. Looking at the value of multiple R-Squared (R^2), which is shown at the top right corner of the screen, we can see that age explains 19.3% of the variation in sexual frequency. (What appears on the screen is the *proportion* of variance explained; to convert it into a *percentage*, simply move the decimal point two places to the right.) The effect of age is significant at the .01 level as the beta value of –.440 is followed by two asterisks. (Note that this

133

beta coefficient is negative, indicating that higher age is associated with lower sexual frequency.)

It is possible that age does not actually have a direct effect on sexual frequency. In general, as people age, their health might decline, and declining health might explain the decline in sexual frequency. Thus, it might be that health is an intervening variable between age and sexual frequency. If so, there might be no relationship between age and sexual frequency when we control for health. We can diagram this argument as follows.

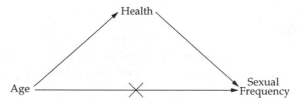

To test this model, let's add health perceptions (HEALTH!) as a second independent variable in the regression analysis.

Data File:	**GSS**
Task:	**Regression**
Dependent Variable:	**122) SEX FREQ!**
➤ *Independent Variables:*	**113) AGE!**
	125) HEALTH!
➤ *View:*	**Graph**

The correlation coefficient of each independent variable with the dependent variable is shown below each line. We can see that health perception is correlated with sexual frequency. The beta coefficient is .104** and is statistically significant at the .01 level. However, the R^2 figure is only slightly higher than it was before: Both independent variables (age and health perception) explain 20.4% of the variation in sexual frequency, but age by itself explained 19.3%. Further, the beta coefficient for age has not changed much from what it was before.

The beta values indicate the effect of one independent variable when the effects of the other independent variables have been controlled. In this situation, the beta coefficient for age (−.415**) means that it has an effect on sexual frequency even when we control for the effects of health. The beta coefficient for health (.104**) means that health perceptions have an effect on sexual frequency even when we control for age. Thus, both independent variables (age and health) have an effect on sexual frequency—although the much higher beta coefficient for age means that it has a greater effect on sexual frequency than health perceptions do. Given that each of these independent variables affects the dependent variable even when controlling for the effects of each other, our previous argument and diagram were incorrect. Based on the results, the diagram should be as follows.

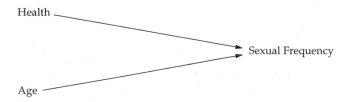

Let's briefly look at the same statistical results in tabular form.

Data File:	**GSS**
Task:	**Regression**
Dependent Variable:	**122) SEX FREQ!**
Independent Variables:	**113) AGE!**
	125) HEALTH!
➤ *View:*	**ANOVA**

Analysis of Variance
Dependent Variable: SEX FREQ!
N: 2312   Missing: 520
Multiple R-Square = 0.204   Y-Intercept = 4.198
LISTWISE deletion (1-tailed test)   Significance Levels: **=.01, *=.05

Source	Sum of Squares	DF	Mean Square	F	Prob.
REGRESSION	1810.677	2	905.339	295.269	0.000
RESIDUAL	7079.736	2309	3.066		
TOTAL	8890.413	2311			

	Unstand.b	Stand.Beta	Std.Err.b	t
AGE!	-0.049	-0.415	0.002	-21.725 **
HEALTH!	0.251	0.104	0.046	5.466 **

If you are continuing from the previous example, simply click the [ANOVA] option.

Only certain parts of this information are important for our purposes. Notice that the number of cases used in this analysis is 2,312 and that 520 cases were missing. All cases that were missing on any variable in this analysis were excluded from the analysis. We can also see from this information that the effects of the two independent variables combined is statistically significant; the entry for the *Prob.* column is the significance level, .000. The effect of each independent variable is statistically significant at the .01 level—there are two asterisks next to the *t* statistic for each.

Now let's try this REGRESSION task with aggregate data. Use the USA data file. In Exercise 2a, we looked at the relationship between hunting and the murder rate and found a significant negative relationship. Murder is an interactive crime that almost always involves contact between the victim and the murderer. This would suggest that murder rates should increase when the amount of contact between individuals increases. For example, during summer months when the weather is pleasant, people spend more time outside and have more contact with one another. In fact, the murder rate is higher in summer than in winter. Areas also differ in their winter weather; in some states, winter is just as pleasant as summer. These states should have higher murder rates than states with extremely cold winters.

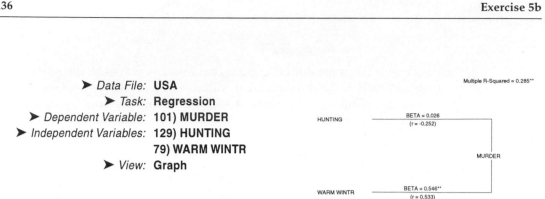

➤ *Data File:* **USA**
➤ *Task:* **Regression**
➤ *Dependent Variable:* **101) MURDER**
➤ *Independent Variables:* **129) HUNTING**
79) WARM WINTR
➤ *View:* **Graph**

The temperature in winter has a strong relationship with the murder rate: r = 0.533. The effect of hunting is no longer significant. Controlling for weather caused a big drop in the effect of hunting on the murder rate. This suggests that the original negative relationship was spurious.

Using the USA file, let's examine another possible spurious relationship.

Data File: **USA**
Task: **Regression**
➤ *Dependent Variable:* **121) VIOLENT CR**
➤ *Independent Variable:* **5) %HISPANIC**
➤ *View:* **Graph**

We can see from these results that there is a positive relationship between %HISPANIC and VIOLENT CR—states that have higher percentages of Hispanics in their population have higher violent crime rates. The beta is .408 and is statistically significant at the .01 level. On this basis, we might come to an incorrect conclusion. In the United States, Hispanics often live in areas that are urban and have substantial poverty, and these are the areas that also have high crime rates. Let's see what happens to the relationship between %HISPANIC and VIOLENT CR when we add two other variables: urbanism and percentage of the state's population below the poverty level.

Data File: **USA**
Task: **Regression**
Dependent Variable: **121) VIOLENT CR**
➤ Independent Variables: **5) %HISPANIC**
6) %URBAN
32) %POV LINE
➤ View: **Graph**

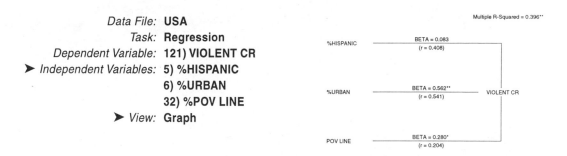

Here we see that the beta for %HISPANIC has been reduced to almost nothing and is no longer statistically significant. Thus, the percentage of the population that is Hispanic has no effect on the violent crime rate when we control for the other two variables, urbanism and poverty. The original relationship between %HISPANIC and VIOLENT CR was spurious. Note that both urbanism and poverty have a statistically significant effect on the violent crime rate even when controlling for the effects of other independent variables in the analysis. The beta for urbanism is strong (.562) and statistically significant at the .01 level. The beta for poverty (.280) is statistically significant at the .05 level. Thus, urbanism is the crucial variable here in explaining the violent crime rate.

Your turn.

Note: In order to challenge you to think about independent and dependent variables, no Software Guides are included in the worksheets for Exercise 5b.

1. Using the GSS data file, test the following causal argument using regression analysis: "The higher their incomes, the more likely people are to support freedom of speech. However, this is really a spurious relationship, since both variables are the result of education—education increases both income and toleration of unpopular speech."

   Diagram this argument using 114) EDUCATION!, 116) R.INCOME!, and 30) COMSPK.

   Use regression analysis to test this causal interpretation. Print the regression graphic. (Note: If your computer is not connected to a printer or if you have been instructed not to use the printer, just skip these printing instructions.)

   a. What is the dependent variable?_____

   b. List the independent variables, and provide the value of beta and the level of significance for each variable.

VARIABLE NAME	BETA	SIGNIFICANCE (Circle one.)
_____	_____	Yes No
_____	_____	Yes No

   c. What is the value of the multiple R^2? _____

d. What is the level of significance of R^2?   _____

e. Is there support of the causal argument that education
   is the antecedent variable and the source of a spurious
   relationship? (Circle one.)   . Yes   No

f. Cite the evidence which supports this conclusion.

2. Using the GSS data file, test the following causal argument using regression
    analysis: "Both father's and mother's education influence how much educa-
    tion their children get."

    Diagram this argument using 118) DAD EDUC!, 119) MOM EDUC!, and
    114) EDUCATION!.

    Use regression analysis to test this causal interpretation. Print the regression
    graphic.

    a. What is the dependent variable?   _____

    b. List the independent variables, and provide the value of beta and the
       level of significance for each variable.

       VARIABLE NAME   BETA   SIGNIFICANCE (Circle one.)

       _____   _____   Yes   No

       _____   _____   Yes   No

    c. What is the value of the multiple R^2?   _____

    d. What is the level of significance of R^2?   _____

e. Is there support of the causal argument that the education of both parents causes the education of their offspring—that is, that each plays an independent causal role? (Circle one.)   Yes   No

f. Cite the evidence which supports this conclusion.

3. Now use 124) PRAYFREQ! as the dependent variable, 113) AGE! as the first independent variable, and 89) SEX as the second independent variable. Diagram the causal form you think these variables would fit. (Note: Sex is being treated as a dummy variable in this analysis, in which females are coded with a higher value than males. Hence, if the variable SEX is positively correlated with another variable, it means that females are positively related with that variable. But if the relationship is negative, it means that females are negatively associated with that variable.)

Use regression analysis to test this causal interpretation. Print the regression graphic.

a. List the independent variables, and provide the value of beta and the level of significance for each variable.

VARIABLE NAME	BETA	SIGNIFICANCE (Circle one.)	
_____	_____	Yes	No
_____	_____	Yes	No

b. What is the value of the multiple R^2?   _____

c. What is the level of significance of R^2?   _____

d. Interpret these findings.

4. Test the following causal argument using regression analysis: "The strength of religious belief affects the frequency of praying. However, the link between strength of religious belief and frequency of prayer is commitment to conventional religious activities such as attendance at religious services."

Diagram this argument.

Use regression analysis to test this causal interpretation. (Hints: Use the search feature to find the variables you need. In selecting your variables, use *uncollapsed* variables—variables that have an ! on the end of their names.) Print the resulting graphic.

a. What is the dependent variable?   _____

b. List the independent variables, and provide the value of beta and the level of significance for each variable.

VARIABLE NAME	BETA	SIGNIFICANCE (Circle one.)
_____	_____	Yes No
_____	_____	Yes No

c. What is the value of the multiple R^2?   _____

d. What is the significance level of R^2?   _____

e. Do the results support the causal argument that conventional religious commitment (as represented by attendance at religious services) is the intervening variable between strength of religious preference and frequency of praying? (Circle one.)   Yes   No

f. Explain your answer, citing the most relevant results.

5. Using the USA data file, look at the effect of variable 18) %COLLEGE on the dependent variable 24) AVG UNEMP$. Write the beta coefficient and correlation coefficient below.

Beta = _____

r = _____

Note that when there is only one independent variable in the regression analysis, the standardized beta coefficient and the correlation coefficient will be the same.

Is this relationship between %COLLEGE and AVG UNEMP$ statistically significant? (Circle one.)   Yes   No

Describe the nature of this relationship between the average unemployment benefit and the percentage of the population with a college degree.

a. Now add an independent variable to this analysis: 31) PER CAP$. List the independent variables, and provide the beta and the level of significance for each variable.

   VARIABLE NAME   STANDARD BETA   SIGNIFICANCE (Circle one.)

   _____   _____   Yes   No

   _____   _____   Yes   No

b. What is the value of the multiple R^2?   _____

c. What is the level of significance of R^2?   _____

d. What happened to the relationship between %COLLEGE and AVG UNEMP$ when you added PER CAP$ to the analysis?

e. Given the results before adding PER CAP$ to the analysis and the results after adding it, diagram what you believe to be the causal connection among these variables.

Selecting a Study Design

OVERVIEW

Different kinds of research questions require different types of research designs. Each type of research (experiments, field studies, surveys, etc.) has advantages and disadvantages. In this exercise, you will gain experience in selecting the appropriate type of research in order to answer a particular type of research question.

BEFORE YOU BEGIN

Please make sure you have read Chapter 6 in the textbook and can answer the following review questions (you need not write any answers):

1. Describe the two fundamental features of experiments.

2. Explain why the experiment is the most powerful research design.

3. Give two reasons social scientists cannot always use experiments.

4. What is the big advantage of survey research, and what is its chief disadvantage?

5. What is field research, and in what situations is it most useful?

6. What is aggregate or comparative research, and how are data collected for such research?

7. Describe the general purpose of content analysis.

The text identifies the following types of research: survey research, comparative research, field research, experimental research, and content analysis. These categories reflect the way in which studies are generally described by social scientists: "Her experiment showed . . . ," "He used content analysis to test the idea that . . . ," "Our survey results suggest that . . . ," and so on. Classifying studies in this way identifies the type of research by its most salient element. Survey research, for example, emphasizes a data collection technique, while content analysis is basically a measurement technique, developing measures by coding content. This is a loose categorization and allows us to examine the elements that are usually combined in each design. In this exercise, we'll look at the types of research questions that are best suited to each design.

These designs differ in the degree of structure required in the research question. Experimental research requires a highly structured question since the manipulation of the independent variable is built into the actual design of the study. Field research requires the least structured question since the researcher is able to observe anything of interest. The other approaches tend to fall between these two extremes. The formulation of the research question, then, is one of the first considerations in selecting a design.

In *exploratory* research, the research question is usually unstructured. Researchers are typically interested in a particular phenomenon and are still searching for connections with other phenomena. *Field* research is frequently used in this situation. For example, a researcher interested in how a new member is socialized into a deviant group might well start by observing how members of such a group interact with a new member or a potential member. When a research question is phrased using "how," field research is probably the best approach. Based on these observations, the researcher might develop some hypotheses on "why" certain approaches are more successful than others or "why" some members are more easily socialized than others.

When a researcher is interested in causes, or why certain phenomena happen, other designs are more appropriate. An *experiment* is the most powerful method for testing a causal hypothesis since the problem of spuriousness is eliminated in the design of the study. So, when testing causal hypotheses, the researcher should first evaluate the possibility of conducting an experiment. Is the independent concept subject to manipulation? Can subjects be assigned randomly to levels of the independent variable? If the independent concept *can* be manipulated and subjects *can* be randomly assigned to levels of the independent variable, then the use of an experiment should be considered. In evaluating social programs, for example, subjects frequently can be assigned to different "treatments," and experimental research is a viable option. Incidentally, while most experiments use individuals as the unit of analysis, other units can be used. For example, a company selling to small businesses might use an experiment to compare the effectiveness of two different marketing approaches. In this study, small businesses would be the unit of analysis and each business would be assigned randomly to one of the two marketing approaches.

The researchers should address two additional concerns before deciding to use an experimental design: ethical questions and feasibility questions. Even if it is possible to manipulate the independent concept, it may not be ethical to do so. For example, one could use an experiment to assess the relative effectiveness of spanking in eliminating undesirable behaviors in children, but randomly assigning children to experimental treatments, one of which involved spanking, raises serious ethical questions. Even when the manipulation of the independent variable raises no ethical concerns, it may not be feasible to conduct an experiment. For example, one could test the effectiveness of a suicide-prevention education program by using an experiment: High school students could be randomly assigned to participate in the program or not. However, this study would not be feasible. Because suicide is a rare phenomenon and very few subjects in either treatment condition would ever commit suicide, an enormous number of subjects would be required to reliably compare the suicide rate across experimental conditions.

If an experiment cannot be used, naturally occurring variation of the independent concept must be measured. With the remaining three designs (survey research, comparative research, and content analysis), the logic of testing causal hypotheses is the same. With each, the researcher must not only examine the relationship between the independent and dependent concepts, but also consider potential sources of spuriousness. So the choice of design hinges on the unit of analysis used in the research question and the nature of the concepts.

If the unit used in the research question is an aggregation, then the best test of the hypothesis would be with data on the relevant unit. With most aggregate units, the researcher can use only data that already have been collected—it wouldn't be feasible to collect data to determine the gross domestic product of each of a set of nations. Lack of appropriate data is frequently a problem in comparative research. With some aggregate units, surveys of "key" persons can be used to supplement existing data. For example, research questions about cities might be addressed by surveying mayors of an appropriate sample of cities. In some cases, it might be possible to use content analysis to create appropriate measures. For example, if you were interested in the effect of legal codes on various types of crime, you might be able to use content analysis of written legal codes to create the independent variable. Or with the Human Area Relations Files, you could use content analysis to code characteristics of societies from field notes recorded by anthropologists.

If the unit of analysis is an individual, then survey research is probably the most appropriate design. Except for ethical issues, the only limits on survey research are that you must be able to identify the appropriate population, to select a sample from this population, if necessary, and to collect information on the relevant concepts by asking questions.

Study designs are not mutually exclusive—elements of two or more may be incorporated into the same study. For example, in the experiment described in the text, subjects are shown information about a political candidate in which gender is the independent variable. Information on the dependent variable, candidate preference, could be collected in a variety of ways. The subjects might be asked to cast a single vote. Or they could be asked to fill out a questionnaire. Or they might be asked to write an essay on the relative merits of the candidates. Content analysis of the essay might be used to develop a measure of the dependent variable.

There are many other ways in which elements of more than one technique might be combined in a single study. A field researcher might survey the members of a group being observed. Or a survey study might have interviewers code information about an individual's living room—a bit of field research. Studies evaluating social programs frequently employ "field experiments"—randomly assigning subjects to one of several experimental treatments conducted in natural settings.

In designing research, the most important task is to examine the research question and to collect data in the manner most appropriate for answering this question.

1. For each of the following research questions, circle the design (experimental, survey, comparative, field, or content analysis) that you think is best and explain why.

   a. On what kinds of issues are there differences between Democrats and Republicans in the general public? (Circle one.)

      Experimental   Survey   Comparative   Field   Content Analysis

      Explain:

   b. Do the popular magazines of today portray minorities differently than they did twenty years ago? (Circle one.)

      Experimental   Survey   Comparative   Field   Content Analysis

      Explain:

   c. Do cities that have higher crime rates have higher rates of unemployment and poverty? (Circle one.)

      Experimental   Survey   Comparative   Field   Content Analysis

      Explain:

d. If an instructor writes encouraging notes to students on their first exams, does this increase student performance on the next exam? (Circle one.)

Experimental   Survey   Comparative   Field   Content Analysis

Explain:

e. What are the similarities and differences between bartenders and barbers in terms of how they handle conversations with customers? (Circle one.)

Experimental   Survey   Comparative   Field   Content Analysis

Explain:

2. For some research designs, it is very important to consider change over time. For each of the following research questions, circle **Yes** or **No** to indicate whether the research would need to use data from at least two different time periods.

a. Do men and women differ from one another in terms of their attitudes toward abortion issues? (Circle one.)   Yes   No

b. Is the percentage of the U.S. population engaged in agriculture increasing or decreasing? (Circle one.)   Yes   No

c. Are nations that are more urbanized more democratic than nations that are less urbanized? (Circle one.)   Yes   No

d. Do members of the American Communist Party hold different views now than they did before the breakup of the Soviet Union? (Circle one.)   Yes   No

e. Are college students today more likely or less likely to major in social sciences than the college students of a decade ago? (Circle one.)   Yes   No

f. Can a government program reduce drug abuse among
    young people? (Circle one.)   Yes   No

g. Are there regional differences in racial attitudes? (Circle one.)   Yes   No

h. Do younger people have different views toward material
    values than older people do? (Circle one.)   Yes   No

i. Are there social attitude differences between those who prefer
    cats as pets and those who prefer dogs as pets? (Circle one.)   Yes   No

3. For each of the following research questions, circle **Yes** or **No** to indicate
    whether it is possible to use an experiment to answer the research question.
    If it is not possible to use an experiment, explain why. If an experiment is
    possible, describe any ethical issue or feasibility problem that might prohibit
    experimentation.

    a. Do males and females hold different views on the issue of pornography?

        NO (Explain below.)

        YES (Describe below any ethical issue or feasibility problem.)

    b. Are people more likely to be persuaded by a speech if it is delivered by
        someone who has an attractive appearance?

        NO (Explain below.)

        YES (Describe below any ethical issue or feasibility problem.)

    c. Are juries more likely to convict an African-American person than a
        white person?

        NO (Explain below.)

        YES (Describe below any ethical issue or feasibility problem.)

d. Are sports fans more aggressive than people who are not sports fans?

NO (Explain below.)

YES (Describe below any ethical issue or feasibility problem.)

e. When election ballots list candidates alphabetically, does this give an advantage to people whose names start with letters that come early in the alphabet?

NO (Explain below.)

YES (Describe below any ethical issue or feasibility problem.)

f. Does political party affiliation make a difference in the voting behavior of state legislators?

NO (Explain below.)

YES (Describe below any ethical issue or feasibility problem.)

4. Different kinds of research questions require different kinds of units of analysis (cases). For example, if you want to explain why some people are more supportive of democratic principles than others are, then you would use individuals as the units of analysis. For each of the following research questions, write in the blank space the unit of analysis for which you would need to collect data.

a. Are the political and social views of people related to the kinds of music they prefer?

_____

b. Do small colleges and universities have higher graduation rates than larger colleges and universities?

_____

c. How is per capita income of states related to their per capita spending on schools?

_____

d. Do increases in economic development within nations lead to lower birth rates?

_____

e. What kinds of social, economic, and political values are reflected in prime-time dramas on television?

_____

f. Does the size of a police department have an effect on how responsive it is to citizen complaints?

_____

g. What kinds of newspapers are most likely to endorse Democratic candidates for public office?

_____

5. Sometimes, we need to combine more than one method in order to investigate research questions. For each of the following situations, circle the *two* methods that we would probably need to use and provide a brief description of how they would be combined.

a. What kinds of people (in terms of personality, social attitudes, political views, and background characteristics) are most likely to be persuaded by an antismoking documentary? (Circle two.)

Experimental Survey Comparative Field Content Analysis

Briefly describe how you would combine these methods.

b. Are there liberal-conservative differences among newsmagazines (*Time*, *Newsweek*, etc.), and are these differences reflected in the views of their readers? (Circle two.)

   Experimental   Survey   Comparative   Field   Content Analysis

   Briefly describe how you would combine these methods.

c. Do the "messages" of restroom-wall graffiti vary according to the type of social environment in which the restroom exists? (Circle two.)

   Experimental   Survey   Comparative   Field   Content Analysis

   Briefly describe how you would combine these methods.

d. Are people who display flags outside their houses on the Fourth of July more supportive of basic democratic principles than are those who do not display flags? (Circle two.)

   Experimental   Survey   Comparative   Field   Content Analysis

   Briefly describe how you would combine these methods.

7

Survey Analysis

OVERVIEW

In this exercise, you'll learn about entering survey data into a computer file, checking for certain types of errors, deciding which responses (e.g., "don't know") to treat as missing, collapsing variables into a smaller set of categories to make them more convenient for cross-tabulation analysis, and recoding variables to create new variables. You will also gain experience in identifying problems in survey research.

BEFORE YOU BEGIN

Please make sure you have read Chapter 7 in the textbook and can answer the following review questions (you need not write any answers):

1. What three main factors cause reliability problems in survey research?

2. Compare the advantages/disadvantages of interviews and questionnaires.

3. Compare the advantages/disadvantages of telephone and face-to-face interviews.

4. What are the benefits of using standard questions in surveys?

5. Compare the advantages/disadvantages of closed and open-ended questions.

6. What problems (e.g., bias) should be considered in writing survey questions?

7. Why are forced option questions better than cafeteria questions?

8. What is a contingency question?

9. In terms of response bias among survey respondents, what are the problems of conformity and response set?

10. How is a trend study different from a panel study?

11. Describe the differences among age, cohort, and period effects.

12. What ethical concerns are particularly relevant to survey research?

The primary challenge in conducting survey research is to ask questions that respondents can and will answer. The text provides considerable guidance in avoiding various problems in question construction, and you will get a chance to evaluate the quality of various questions in the written exercises. However, even after the data have been collected, there is still a great deal of work to do before you can analyze the data.

First, with the exception of some situations in which data are entered as the survey is being done (e.g., in computer assisted telephone interviews, or CATI), we need to enter the data from the survey into a computer file. Second, we need to prepare the data for analysis. This involves checking the accuracy of the data, identifying inappropriate responses, and eliminating unwanted categories. We might also need to *collapse* variables (e.g., collapsing education in actual years to a set of categories such as *under 8 years*, *8–11 years*, etc.), and we might also need to *recode* variables to make them more useful for analysis (e.g., combining responses from several questions concerning participation in social activities to create a social participation index).

DATA ENTRY

The following illustrations show three questions from two hypothetical question-naires filled out by two respondents. For each question, respondents were instructed to circle the number of the alternative that most reflected their answers. Thus, this questionnaire is *pre-coded*. That is, numbers—codes—have already been assigned to the question alternatives. If numbers had not already been assigned to the possible responses, then we would need to go through the questionnaires and code the responses.

Respondent 1

Sex: (1) Male   2. Female

During the last twelve months, have you written a letter to any public official to express your views on any matter?

(1) Yes   2. No   3. Don't Know

Please indicate your level of agreement or disagreement with the following state-ment: Judges should deal more harshly with criminals convicted of violent crimes.

1. Strongly Agree   (2) Agree   3. Disagree   4. Strongly Disagree   5. Don't Know

---

```
┌─────────────────────────────────────────────────────────────────────────┐
│ Respondent 2                                                              │
│                                                                           │
│   Sex:    1. Male      ②  Female                                          │
│   During the last twelve months, have you written a letter to any public │
│   official to express your views on any matter?                          │
│   1. Yes      2. No     ③  Don't Know                                     │
│   Please indicate your level of agreement or disagreement with the        │
│   following statement: Judges should deal more harshly with criminals     │
│   convicted of violent crimes.                                            │
│   1. Strongly Agree   2. Agree   3. Disagree  ④ Strongly Disagree  5. Don't Know │
│                                                                           │
└─────────────────────────────────────────────────────────────────────────┘
```

Next we need to organize the data prior to entering it into a computer file. The data are usually organized in a spreadsheet fashion with cases (respondents) along the side and variables across the top. For example, if we take the two hypothetical respondents and three hypothetical variables in the preceding illustrations, we would organize these data as follows:

	Variable 1 SEX	Variable 2 WRITE LET	Variable 3 HARSHER?
Respondent 1	1	1	2
Respondent 2	2	3	4

You have seen in the MicroCase data files that each variable has a number and a name. Here we named the first variable SEX, the second variable is WRITE LET, and the third variable is HARSHER?. Note that for SEX, the first respondent is coded 1 (for male). For WRITE LET, respondent 1 is coded 1 (for Yes). For HARSHER?, respondent 1 is coded 2 (for Agree). Respondent 2 is coded 2 (Female) for SEX, 3 (Don't Know) for WRITE LET, and 4 (Strongly Disagree) for HARSHER?.

Having organized the data, you would now be ready to create a data file and enter the data. Thus, at this point you might want to go to "Creating a MicroCase File" at the beginning of the "Projects" section of this book and go through the brief tutorial on creating a data file. Alternately, your instructor might provide a separate data entry exercise for you to do.

Unfortunately, errors can occur in the process of gathering and entering data, and we will discuss ways of identifying certain types of errors in the next section. However, at the data entry stage, researchers frequently use *rekey verification* in order to minimize or eliminate the errors that can occur at this stage. Rekey verification (which is included in your Student MicroCase) means that the data are entered a second time and checked against the original data entries. If the second entry does not match the first entry, then the program notifies you so that you can see which entry is correct. This checking technique is based on the assumption that a data entry error is unlikely to occur a second time on the same response. You may remember that this is the same basic idea used to construct measures of reliability. By rekeying the data, we increase the reliability of the data.

To take a hypothetical example of rekey verification, when someone entered data for respondent number 800 for a religious preference variable, a 3 was entered. However, in doing rekey verification for this respondent for this variable, a 2 was entered. The rekey verification program will immediately respond that there is a discrepancy and ask which of the two values (the 3 or the 2) is correct. At this point, the person entering the data would determine which number is correct and select that number.

PREPARING DATA FOR ANALYSIS

Even after the survey data have been entered into a data file, there is still work to be done before we are ready to analyze the data.

CHECK DATA FOR ACCURACY AND INTEGRITY

After the data file has been created, we need to check the data entries to make sure that they are correct. Rekey verification (if it is available) can minimize errors at the data entry stage, but some data entry errors can still occur. Further, problems can occur prior to data entry. For example, a respondent's answer to a question might have been coded incorrectly. Another problem is that some respondents do not take the survey seriously and might give wrong answers as a joke or for other reasons. Such problems can occur regardless of whether the researcher creates the data file, obtains it from someone else, or imports the file from another program.

Beyond rekey verification, we can check data for accuracy by examining the data entries for certain kinds of problems. We begin this process by simply looking at the data to see whether everything looks right. Sometimes this can identify problems. In order to demonstrate this process, we have created a small fictitious TEST file. Let's look at the data in this file.

> ➤ Data File: **TEST**
> ➤ Task: **List Data**

Scroll across the screen so that you can view all the data. One problem that can occur is that one or more cases might not have the right values for variables entered in the proper columns. This can occur, for example, when the person entering the data gets off onto the wrong column—although MicroCase makes it very unlikely that errors of this sort will occur. Also, it is not unusual for something like this to happen when a researcher imports data from one program into another program. Thus, we first need to check to make sure that all the rows and columns end at the same point. If, for example, a row ended several columns before other rows did, then there is very likely a problem with the data for this case and it would need to be checked. As you can see, in the TEST file, all rows and columns end at the same point. Thus, the file looks good so far.

While all the rows and columns end at the appropriate points, this does not guarantee that the right data are in the right places. So, it would be a good idea to check some of the data for a couple of variables to make sure that the data look right. For example, we might check some of the cases for the last variable. If the last variable is correct, this indicates that the other variables are very likely lined up properly.

The next step is to examine the univariate statistics for each of the variables to determine whether anything looks like a mistake. Let's start with the first variable, SEX.

Data File: **TEST**
➤ *Task:* **Univariate**
➤ *Primary Variable:* **1) SEX**
➤ *View:* **Bar - Freq.**

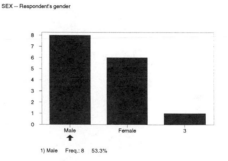

Do you notice anything wrong here? Males are coded 1 and females are coded 2, but one case is coded 3. Thus, there is a data entry error, and we would need to check this case and correct the error. Since we have found one error for this case, it would be a good idea to check a few other variables for this case to make sure that this is just an isolated error rather than part of a pattern. Actually, this kind of data entry error would not occur in MicroCase because you can set the range for each categorical variable. Thus, the variable would have a low value of 1 and a high value of 2. If you tried to enter a 3 for this variable, the program would alert you that the value is outside the range of allowed values. Thus, using MicroCase, we had to "cheat" in order to even demonstrate this problem. However, not all statistical analysis systems have the capability to check the range.

We put in two more of these "errors" in this fictitious data file to demonstrate this kind of error. You might try finding them.

IDENTIFY INAPPROPRIATE RESPONSES

Sometimes, respondents give responses that are not accurate. For example, some survey respondents give joking responses that do not reflect their actual situations. We need to try to identify such responses, especially when they are extreme answers. There are several ways to check for such responses. When the values of a variable are not limited to particular categories, you should look at the univariate distribution to see if there are any extreme values. For example, in one year of the GSS, one individual claimed a lifetime total of 403 female sexual partners. This seemed highly unlikely when compared to the rest of the sample. But closer examination of this case revealed that the respondent was a 69-year-old man who reported having 5 to 10 sexual partners during the previous 12 months. He claimed casual pickups and paid sexual partners in addition to his wife. Assuming this man had been similarly active since the age of 20, his claim of 403 partners over his lifetime seems reasonable, even if the precision of the answer seems rather startling.

In the fictitious TEST file, let's look at the ATTEND variable.

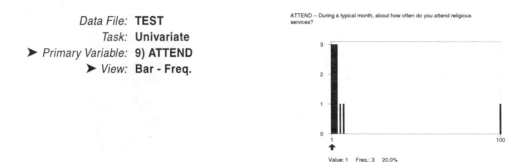

Note that for all but one of the respondents the answers given to this religious attendance question range from 1 to 8 times per month. The remaining respondent, however, indicated that he or she attended religious services 100 times per month. While this figure sounds very high and the roundness of the number makes it suspect, it is possible that it is actually correct. So, what do we do?

There are three options when we identify a suspicious value. First, we could change that value to missing data. Second, we could eliminate that case completely—if the respondent's answer to one question is a distortion, then the respondent's answers to other questions might also be unreliable. Third, we could create a "flag" variable. To use a "flag" variable, you first create a new variable with the values 0 and 1 (or actually any values that you wish). If a case appears to have inappropriate data, you set the value of this flag variable to 1; otherwise, the case is given a value of 0. When you analyze the data, you always select a subset of those cases that have a 0 on this flag variable. The advantage of using a flag variable rather than eliminating cases is that the data still exist for other analyses or other researchers.

Let's see how this flag variable option works. Note that the last variable in the TEST file is named FLAG VAR9—it is a flag variable for variable 9) ATTEND. For ATTEND, the only case that seemed unusual was the one that claimed to attend religious services 100 times a month. This case has been given a value of 1 for the flag variable and all other cases have a 0. So, we could use this flag variable to exclude any cases that are coded 1. Let's do this.

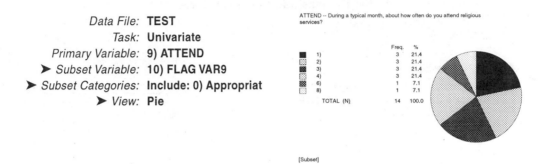

Data File:	**TEST**
Task:	**Univariate**
Primary Variable:	**9) ATTEND**
➤ *Subset Variable:*	**10) FLAG VAR9**
➤ *Subset Categories:*	**Include: 0) Appropriat**
➤ *View:*	**Pie**

ATTEND -- During a typical month, about how often do you attend religious services?

		Freq.	%
■	1)	3	21.4
	2)	3	21.4
■	3)	3	21.4
	4)	3	21.4
	6)	1	7.1
	8)	1	7.1
	TOTAL (N)	14	100.0

[Subset]

Note that the exclusion of this case for this variable has quite an impact on the results. However, in a typical survey of 1,500 or so respondents, one highly unusual case such as this is not likely to have a great impact on the results.

Another way to check for inappropriate responses is to examine the associations between variables known to have certain relationships. For example, in a college survey, students may be asked their sex and their place of residence. We know that anyone living in a fraternity should be male. If we find one or more females claiming to live in a fraternity, we probably would want to check their responses to other questions. In a high school survey, one student claimed to be 7 feet tall and weigh 100 pounds. This is highly unlikely. Let's see how such checks might be done using the TEST file.

Data File:	**TEST**
➤ *Task:*	**Cross-tabulation**
➤ *Row Variable:*	**8) VOTE96**
➤ *Column Variable:*	**7) REGISTERED**
➤ *View:*	**Table**
➤ *Display:*	**Column %**

7) REGISTERED

8) VOTE96	Yes	No	Don't know
Clinton	41.7%	50.0%	0.0%
Dole	50.0%	0.0%	0.0%
Perot	8.3%	0.0%	0.0%
Other	0.0%	50.0%	0.0%
No Answer	0.0%	0.0%	100.0%
TOTAL	100.0%	100.0%	100.0%

V=0.870**

Note that two respondents who said they were not registered to vote in the 1996 elections also claimed to have voted—one for Clinton and one for Other. Thus, we know that for these two respondents, there are inappropriate responses somewhere. Either they were actually registered or they did not actually vote.

ELIMINATE UNWANTED CATEGORIES

Note that the preceding cross-tabulation table (VOTE96 by REGISTERED) contained several categories that are not very useful for analysis in this situation. Another aspect of cleaning survey data for analysis involves answers such as "don't know," "other," and "no answer." Generally, we want these answers to be treated as missing data during analysis. In MicroCase, the easiest method of doing this is to list these as additional categories of missing data. Let's see how this works.

> ➤ *Data File:* **GSS**
> ➤ *Task:* **Cross-tabulation**
> ➤ *Row Variable:* **10) EVER STRAY**
> ➤ *Column Variable:* **89) SEX**
> ➤ *View:* **Table**
> ➤ *Display:* **Column %**

EVER STRAY   by   SEX
Cramer's V: 0.111 **

		SEX		
		Male	Female	TOTAL
E V E R S T R A Y	Yes	169	155	324
		15.8%	11.2%	13.2%
	No	581	900	1481
		54.3%	65.1%	60.4%
	Never wed	291	296	587
		27.2%	21.4%	23.9%
	No Answer	28	31	59
		2.6%	2.2%	2.4%
	Missing	163	218	381
	TOTAL	1069	1382	2451
		100.0%	100.0%	

If you read across the row labeled "No answer," you will see that there are a total of 59 respondents who did not answer this question. In Exercise 5a of this workbook, you used a method to *temporarily combine* categories in a cross-tabulation table. This same technique can be used to *temporarily drop* a category. To do this, click on the category label "No answer," which causes the entire row to become highlighted. Then click on the [Collapse] button and choose the option to convert this category to missing data. Click [OK] and the table returns to the screen. Notice that the "No answer" category has been temporarily removed. (If you were to repeat the cross-tabulation of these same two variables, the "No answer" category would reappear again.)

Student MicroCase has a second way to temporarily turn categories to missing data. Let's say you have a data set that has a large number of variables that include a "No answer" category—or some other unwanted category, such as "Don't know" or "No opinion." Student MicroCase allows you to designate a set of such categories so that whenever they are encountered during analysis they will automatically be turned to missing data. Return to the **FILE & DATA MENU** and select the [File Settings] option. In the lower right corner of the [File Settings] box, you can enter up to six missing data categories that will be excluded from all subsequent analyses while using this data file. Go ahead and type **NO ANSWER** in the first blank. This category will now be excluded not only from EVER STRAY, but also from all variables in all analyses that have this category in the GSS file. You could also enter DON'T KNOW or NO OPINION and these categories would be excluded from our analyses.

If we later change our minds and want to include any of these categories again, we can go back to this option and delete the category from the list. In your student version of the software, all categories you enter here will be erased when you open a different data file or exit the program.

Now that you have "No answer" assigned as a missing data category, return to the CROSS-TABULATION task and repeat your analysis of EVER STRAY and SEX.

Data File: **GSS**
➤ *Task:* **Cross-tabulation**
➤ *Row Variable:* **10) EVER STRAY**
➤ *Column Variable:* **89) SEX**
➤ *View:* **Table**
➤ *Display:* **Column %**

EVER STRAY   by   SEXCategories set to Missing Data (in File Settings):
NO ANSWER

Cramer's V: 0.111 **

		SEX		
		Male	Female	TOTAL
	Yes	169	155	324
		16.2%	11.5%	13.5%
	No	581	900	1481
		55.8%	66.6%	61.9%
	Never wed	291	296	587
		28.0%	21.9%	24.5%
	Missing	191	249	440
	TOTAL	1041	1351	2392
		100.0%	100.0%	

Sure enough, the "No answer" category was automatically turned to missing data.

COLLAPSING VARIABLES

Previously in this workbook, you saw the usefulness of using collapsed versions of a variable. You may recall, for example, that all the variables in the data set that end with a "!" are the original, uncollapsed version of that variable. Here is the univariate distribution of the original age variable in GSS:

Data File: **GSS**
➤ *Task:* **Univariate**
➤ *Primary Variable:* **113) AGE!**
➤ *View:* **Bar - Freq.**

AGE! -- RESPONDENT'S AGE (AGE)

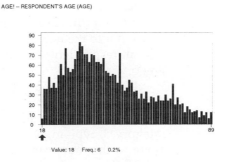

Value: 18   Freq.: 6   0.2%

Notice that the original age categories range from 18 to 89. If we tried to use this variable in a cross-tabulation analysis, the resulting table would be nearly impossible to interpret. If you look at variable 101) AGE, you'll see that this variable has been collapsed into five categories.

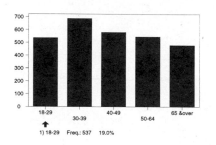

AGE -- RESPONDENT'S AGE GROUP (AGE)

You may have wondered how the collapsed version of age was created. Did someone go through the data set and assign each case into one of these five categories and then key in the data for the new variable? In the old days that's how this was done. But today there are powerful programs like MicroCase that make this an easy procedure.

Go to the **FILE & DATA MENU** and select the COLLAPSE VARIABLE task. You are prompted to select a variable to collapse. Even though a collapsed version of AGE! already exists in this data set, let's create another one ourselves. So select **113) AGE!** as the variable to be collapsed, and click [OK]. This bar graph is similar to what we saw moments ago when we used the UNIVARIATE task to look at this variable. But in this task, we are able to combine specified values on the bar graph. In the lower, middle section of the screen is a list of buttons for categories of the new variable we are about to create. Click on Category 1 and type in **18–29**. Now position your mouse over the first category in the upper bar graph, which represents the respondents who are 18. Notice that there are 6 cases (or 0.2%) in this category (these values are shown below the bar graph). Click on this category to assign it to the new 18–29 category you created below. Click on the next 11 bars until all respondents coded in the 18–29 range have been assigned to this 18–29 category.

Now click on the Category 2 button in the list below and type **30–39**. Return to the upper bar graph and click on the 10 bars representing those respondents who are 30–39 so that they will be assigned to the 30–39 category you created. For Category 3, type in **40–49** and assign the appropriate categories from the graph above to this new category. You're probably getting the hang of this by now. Follow the same procedure for those who are 50–64 and for those who are 65 & over.

Recall that in the previous section we temporarily dropped categories from our analysis. However, if we wanted to permanently drop one of these categories, we could do so by simply not assigning it to one of the categories. Any value in the original variable that is not assigned to a category on the new variable will become missing data and will be automatically excluded from analyses when the new variable is used. In this example, we wanted to assign all of the categories from AGE! to our new variable, so all the categories have been assigned.

As you probably noticed, as we collapsed variables, the horizontal bar graph in the lower right kept a running total of the size of each new category. This helps us balance the sizes of categories if that's an important consideration. Note that we could click the [Cumulative] button on the left if we wanted to work with a cumulative bar graph instead of a regular bar graph. A cumulative graph is often helpful with interval/ratio or ordinal variables in assigning given proportions of cases to new categories. For example, we might want to divide the age distribution into thirds.

We are nearly finished creating our new variable. Click the [Finish] button and a final window will appear on the screen. MicroCase's COLLAPSE VARIABLE option does not replace the original variable but, rather, creates a new variable and places it at the end of the data file. At this point we can enter a name for our new variable. MicroCase suggests just adding a 2 to the old variable name—AGE!2. Let's instead replace that suggestion with the name AGE CATEG, which will remind us that this is an age variable that has been collapsed into categories. If you want to modify your variable description, you can do that as well. Again, MicroCase has suggested some wording for the new variable description, but you may want to modify this to your liking. Click [OK] and you will be told that a new variable has been added to your GSS data file. If you click **[OK]** again, you will be returned to the main menu. The AGE CATEG variable will remain for you to use unless you replace it with another variable. Student MicroCase allows you to add 20 variables like this to your data file. If necessary, new collapsed variables can replace any of these 20 variables.

Let's take a look at the new variable we created.

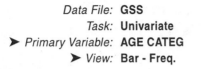

Data File: **GSS**

Task: **Univariate**

➤ Primary Variable: **AGE CATEG**

➤ View: **Bar - Freq.**

AGE CATEG -- Collapsed from AGE!: RESPONDENT'S AGE (AGE)

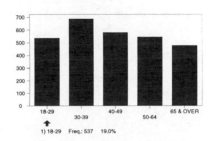

Sure enough. This looks identical to the 101) AGE variable that had already been collapsed in the GSS data file.

RECODING VARIABLES

The collapse option discussed in the previous section is actually one of many methods for creating *recoded variables*. Recoded variables are transformations or computations made to the original set of variables in order to create entirely new variables. There are many different types of recodes used in survey research. One of the more common recodes is to create an *index* that combines responses from two or more questions.

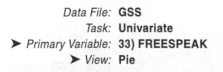

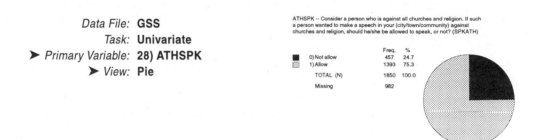

Notice that this variable is coded 0 for respondents who would not allow free speech in this situation and it is coded 1 for those who would allow free speech. If you examine the variable descriptions for the next four variables in the data set, 29–32, you will see that the same coding scheme (0=No; 1=Yes) is used. We could compute a new variable that is based on the summing of the number of "allow speech" responses that each respondent gave to the five freedom of speech questions. In fact, variable 33) FREESPEAK has done just that.

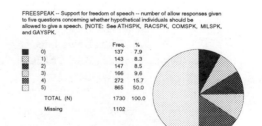

As you can see, those respondents who answered Yes to each of the five free speech questions are coded as 5. Those respondents who answered Yes to all but one of the questions are coded as 4, and so on. (If a respondent did not provide an answer for each of the five questions, they were assigned to missing data for this variable.)

You are probably wondering how we computed this index. Go to the **FILE & DATA MENU** and select the RECODE VARIABLES task. Then select the [Sum/Index] option. The screen that appears gives you numerous choices for the creation of your recode. Although you are welcome to read the various options, do not change the original, default settings for this example. Click the [Next] button to continue. Now select variables **28–32** as the variables to be included in the sum recode. Click the [Next] button and you are presented with a screen that lets you evaluate the reliability of the index you are creating. At the lower part of the screen, you are shown the correlations between each of the variables you selected for the index. At the top of the screen is Cronbach's alpha. You will recall from Chapter 3 in the textbook that Cronbach's alpha is a popular measure of reliability based on the correlation—the internal consistency—of the variables that comprise an index. The rule of thumb is that if Cronbach's alpha is over .70, the measure is sufficiently reliable. Since the value we obtained is .83, we seem to be on safe ground.

Click the [Next] button and you are prompted for the variable name of the index you are creating. Let's call this variable **SPEAK INDX**. If you would like to modify the variable description that MicroCase has automatically generated, you can do that at this time too. Before you continue, take a quick look at the matrix provided at the bottom of this screen. The first column, shown in red, lists the individual values for the new variable you are creating. The next five columns show the coding for the original five variables selected for the index. Examine the first case. Notice that it is coded as 3 for the new variable. If you scroll across and look at the coding of this case on the other five variables, you will see, as expected, that this person answered Yes (coded as 1) to three of the free speech questions. If you go back and examine the sixth case, you'll see that it is coded as 5 in our new variable and that this person answered Yes to each of the five free speech questions. Note that some cases do not have a score. Only those respondents who answered all five questions are included in the index.

After you are done examining this screen, click [Next] to return to the final screen. Here you are simply reminded that your new variable, SPEAK INDX, will be added to the end of the data file. Click [Finish] and [OK] to return to the main menu.

Let's look at the final results of the variable we created.

Data File: **GSS**

Task: **Univariate**

➤ Primary Variable: **SPEAK INDX**

➤ View: **Pie**

SPEAK INDX -- Sum of the following variables: ATHSPK, RACSPK, COMSPK, MILSPK, GAYSPK -- Cronbach's alpha is 0.83 (listwise deletion). There are 1102 missing cases.

		Freq.	%
■	0)	137	7.9
▨	1)	143	8.3
■	2)	147	8.5
▨	3)	166	9.6
▨	4)	272	15.7
▨	5)	865	50.0
	TOTAL (N)	1730	100.0
	Missing	1102	

No surprises here. The results match the 33) FREESPEAK variable we looked at earlier.

1. Surveys are extremely useful in social research, but we need to interpret the results with some caution. Let's look at reported voting participation in the 1996 elections.

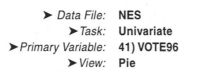

➤ *Data File:* **NES**
➤ *Task:* **Univariate**
➤ *Primary Variable:* **41) VOTE96**
➤ *View:* **Pie**

a. According to the univariate statistics, what percentage of those who answered this question claimed that they voted in the 1996 elections? _____%

b. We know from voting statistics that only about 50 percent of the electorate voted for president in the 1996 elections. What might account for the difference between this figure and the percentage who claimed they voted?

c. Take another look at the variable description for 41) VOTE96. In this NES survey, people were also asked whether they voted in the 1998 congressional elections. Look at the wording of the question in 42) VOTED98?.

Which of the two questions (VOTE96 or VOTE98?) do you think would produce the most accurate responses from people concerning whether they actually voted? (Circle one.)   VOTE96   VOTE98?

Explain your answer.

2. Let's look at 43) PRESVOTE96.

    Data File: **NES**
    Task: **Univariate**
    ➤ *Primary Variable:* **43) PRESVOTE96**
    ➤ *View:* **Pie**

    a. What percentage of the respondents said that they voted
        for Clinton?   _____%

    b. On the basis of voting statistics, we know that President Clinton received
        just under 50 percent of the votes in 1996. Thus, there is a discrepancy
        between the percentage who say they voted for him and the actual per-
        centage. There are at least two basic explanations for this discrepancy.
        One is sampling error. What is another possible explanation?

3. Let's see how the way in which we handle missing data can affect relation-
    ships. Using the NES data file, let's examine the question of how partisan-
    ship affected people's views of who deserved credit for the federal budget
    surplus in 1998. We can hypothesize that Democrats in the general public
    would be more likely to give President Clinton credit, and Republicans
    would be more likely to give the Republican-controlled Congress the credit.
    We will begin this examination by using two variables that still have missing
    values included in them; the exclamation marks at the end of these variable
    names indicate that these variables have not been collapsed at all.

    Data File: **NES**
    ➤ *Task:* **Cross-tabulation**
    ➤ *Row Variable:* **113) SURPLUS!**
    ➤ *Column Variable:* **114) POLPARTY!**
    ➤ *View:* **Tables**
    ➤ *Display:* **Column %**

    a. Look at the column percentages in the table. Then obtain Cramer's V and
        the significance level and write them below.

V = _____

Prob. = _____

b. This table includes categories from both variables that should definitely be treated as missing here (e.g., the "Don't Know" and "No Answer" categories). Further, given our research question, there are other categories that should be treated as missing. Rather than using the collapse or recode techniques discussed earlier, we will use the subset option to include certain categories and exclude others.

Since we are primarily interested in whether Democrats and Republicans differ from one another in giving credit for the budget surplus, let's just include the Democrats and the Republicans from the POLPARTY! variable. For the SURPLUS! variable, let's include just the following three categories: Clinton, Both, and Republican Congress.

Note that in the following software guide, you are selecting two subset variables.

Data File:	**NES**
Task:	**Cross-tabulation**
Row Variable:	**113) SURPLUS!**
Column Variable:	**114) POLPARTY!**
➤ Subset Variable:	**113) SURPLUS!**
➤ Subset Categories:	**Include 0) Clinton**
	1) Both
	3) Rep. Congr
➤ Subset Variable:	**114) POLPARTY!**
➤ Subset Categories:	**Include 1) Democrat**
	3) Republican
➤ View:	**Tables**
➤ Display:	**Column %**

Fill in the column percentages in the following table and then obtain Cramer's V and the significance level and write them in.

	Democrats	Republicans
Clinton	_____	_____
Both	_____	_____
Repub. Congr.	_____	_____

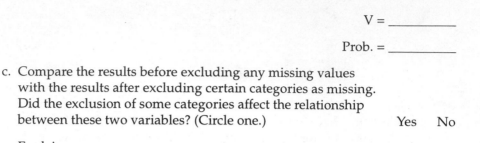

c. Compare the results before excluding any missing values
   with the results after excluding certain categories as missing.
   Did the exclusion of some categories affect the relationship
   between these two variables? (Circle one.)   Yes   No

   Explain your answer.

d. The hypothesis was that Democrats would give Clinton credit for the
   budget surplus and Republicans would give the Republican-controlled
   Congress credit.

   Did the results support the hypothesis? (Circle one.)   Yes   No

   Explain your answer.

4. Based on what you learned from the textbook, discuss the problems with
   each of the following questions.

   a. *Question: Do you think that high schools should or
      should not be prohibited from banning the wearing of
      gang colors by students? (Circle one.)*   *Should   Should not*

   What is the major problem with this question?

   How could this question be rewritten to eliminate the problem?

b. *Question: Should judges be allowed to continue coddling criminals by giving some of them suspended sentences for the first offense?* (Circle one.)   *Yes*   *No*

What is the major problem with this question?

How could this question be rewritten to eliminate the problem?

c. *Question: What is your view concerning abortion?* (Circle one.)

   A. It should not be allowed.
   B. It should be allowed.

What is the major problem with this question?

How could this question be rewritten to eliminate the problem?

d. *Question: How many times have you changed your major since you began college?* (Circle one.)

          0–2 times   *3–5 times*   *More than 5 times*

What is the major problem with this question?

How could this question be rewritten to eliminate the problem?

e. *Question: During the last twelve months, have you done any of the following?* (Circle all that apply.)

Visited an art museum?
Attended a professional sports event?
Gone to a play at a theater?
Gone to an auto race?
Gone to an amateur sports event?

What is the major problem with this question?

How could this question be rewritten to eliminate the problem?

5.   Open the GSS data set and do the following analysis.

> ➤ *Data File:* **GSS**
> ➤ *Task:* **Univariate**
> ➤ *Primary Variable:* **10) EVER STRAY**
> ➤ *View:* **Pie**

a. What is the variable description of 10) EVER STRAY?

_____

_____

_____

b. Fill in the category labels and percentages for this variable.

CATEGORY LABEL   PERCENTAGE

1. _____   _____

2. _____   _____

3. _____   _____

4. _____   _____

c. Which category should be treated as missing data in any analysis?

_____

d. Why are those who have never married placed in a separate category?

e. In the next example, we want to exclude (i.e., turn to missing data) certain categories from the variables we select. Again, rather than use the collapse or recode techniques described in the preliminary part of this exercise, we are going to use the subset option to achieve the same results.

Let's look at the relationship between 10) EVER STRAY and age for all respondents who answered the question about straying. Use column percentages and fill in the following table.

> Data File: **GSS**
> ➤ Task: **Cross-tabulation**
> ➤ Row Variable: **10) EVER STRAY**
> ➤ Column Variable: **102) OVER 50?**
> ➤ Subset Variable: **10) EVER STRAY**
> ➤ Subset Categories: **Exclude 4) No Answer**
> ➤ View: **Tables**
> ➤ Display: **Column %**

Remember, subset variables are selected at the same time that other variables are selected. In this example, make sure you EXCLUDE the No Answer category.

	Under 50	50 & Over
Yes	_____	_____
No	_____	_____
Never Wed	_____	_____

V = _____

Prob. = _____

f. Now let's exclude both the No Answer responses and the Never Wed responses. Then use column percentages to fill in the following table.

Data File:	**GSS**
Task:	**Cross-tabulation**
Row Variable:	**10) EVER STRAY**
Column Variable:	**102) OVER 50?**
Subset Variable:	**10) EVER STRAY**
➤ *Subset Categories:*	**Exclude 4) No Answer**
	3) Never Wed
➤ *View:*	**Tables**
➤ *Display:*	**Column %**

Perhaps the easiest way to modify the category selection for a subset variable is to delete the subset variable entirely and then select it again. For this example, make sure you EXCLUDE the No Answer and Never Wed categories.

	Under 50	50 & Over
Yes	_____	_____
No	_____	_____

V = _____

Prob. = _____

g. Explain which table would be more appropriate for testing the hypothesis that individuals 50 and over are more likely to have strayed than are individuals under 50.

h. How was the strength of the relationship affected by which categories were treated as missing data?

6. The following paragraph describes a study conducted in 1935:

   *Long and Share Our Wealth stimulated the first scientific public opin-
   ion poll on a Presidential race. . . . Emil Hurja, the chief statistician and
   an executive director of the Democratic National Committee, mailed
   straw ballots and a cover letter on April 30, 1935 to about 150,000
   people in the U.S. states. He made the poll appear as if it were being conducted by a magazine, the nonexistent National Inquirer. All the information was to be received from a three-by-five inch card—the two-cents
   postage was prepaid.*[1]

   What element of this study might raise ethical questions? Explain.

7. Let's compute an index. Note that variable 55) ABORT INDX in the GSS data
   file is an index based on the number of "allow abortion" responses given to
   six questions asking about abortion in various situations. Let's compute an
   index based on just three of these questions—those that concern emergency
   situations. The questions ask whether the respondent agrees that abortion is
   acceptable when the health of the mother is endangered, when there might
   be serious damage to the fetus, or when the pregnancy occurred because of
   rape. (If you have difficulty using this task, refer back to the last part of the
   preliminary section of this exercise.)

Data File:	**GSS**
➤ Task:	**Recode Variables (located on the FILE & DATA MENU)**
➤ Recode Option:	**Sum/Index**
➤ Select Variables:	**49) ABORT DEF**
	51) ABORT HLTH
	53) ABORT RAPE

   After you have selected these variables for the index, click on [Next].

---

[1] Edwin Amenta, Kathleen Dunleavy, and Mary Bernstein, "Huey Long's 'Share Our Wealth' and the
Second New Deal," *American Sociological Review*, Oct. 1994.

a. What is the value for Cronbach's alpha?   _____

b. According to the textbook, is this value high enough to
   be reliable? (Circle one.)   Yes   No

Click [Next] again. The program presents the value for this new variable for each respondent. Note also that there are many missing values for this variable—because not all the respondents were asked this series of questions and some of those who were asked did not answer.

Scroll through the cases and write in the value of this new variable for the following cases. (Note: Write in *Missing* for a case if it has a blank.)

Case Number 1   _____

Case Number 28   _____

Case Number 31   _____

Case Number 42   _____

Now type a name for this new variable—let's call it **ABORT EMER**—and press <ENTER>. This index is complete.

Go to the Univariate task and examine the variable you created.

Data File:	**GSS**
➤ *Task:*	**Univariate**
➤ *Primary Variable:*	**ABORT EMER**
➤ *View:*	**Pie**

Print out this result and turn it in with your assignment.

Comparative Methods

OVERVIEW

In this exercise, you will first learn more about selecting the proper base for the calculation of rates (e.g., average yearly food stamp benefit per recipient) in comparative research. Because comparative research often uses a relatively small number of cases, each individual case might have a substantial effect on the results. Thus, you will see the effect that outliers can have on correlations and how this problem can be handled. Lastly, you will see that missing data can be a major problem in comparative research.

BEFORE YOU BEGIN

Please make sure you have read Chapter 8 in the textbook and can answer the following review questions (you need not write any answers):

1. In comparative research, describe what a rate is and give examples of rates that have different kinds of bases.

2. Compare the reliability of aggregate data with that of survey data, and explain why there is a difference.

3. What are the chief limitations of aggregate data based on official records?

4. If a rate such as a crime rate is not accurate, how can it still be useful for comparative research?

5. What is the difference between a case-oriented approach and variable-oriented research?

6. What are outliers, and why is it important to check for their existence in comparative research results?

7. Why shouldn't you use 20 independent variables in a regression analysis to explain variation in welfare spending among the states of the United States?

8. What are the primary kinds of circumstances in which ethical issues need to be considered in comparative research?

In this exercise, we'll explore some of the special problems of comparative research. First, choosing an appropriate base (or denominator) for a rate can be difficult, and different bases may lead to quite different results. Second, there are usually a limited number of cases in the data file, so the impact of each case is considerably greater than in survey research. Finally, there may be a sizable amount of missing data.

In comparative research, almost all the variables are rates of one type or another. In selecting the base (or denominator), you always should consider the concept you are trying to measure. In some situations, changing the base helps improve the validity of the measure. For example, dividing the number of students by the *school-age* population would be a better measure of current participation in the educational system than would dividing the number of students by the *total* population. If we used the total population to create this rate, then a country with a small number of school-age children would have a low rate, even if every child was enrolled in school. The difference between the crude birth rate and the fertility rate described in the textbook is another example of refining a rate by changing the base.

Sometimes, however, changing the base may completely change the concept being measured. For example, if you divide the number of female students by the number of females in the school-age population, you will obtain the proportion of *school-age females* who are enrolled in school. On the other hand, if you divide the number of female students by the total number of students, you will get the proportion of *students* who are female. Clearly, these are not the same, and countries would not necessarily rank in the same order on the two variables. Countries in which only the elite enrolled in school but that had no sex discrimination would have a relatively low proportion of school-age females enrolled in school, but a relatively high proportion of students would be female.

Rates may change (or differ) because the values of either the denominator or the numerator or both change. Consider, for example, the percentage of deaths between ages 15 and 25 attributable to suicide. The numerator is the number of deaths by suicide in this age group, and the denominator is the total number of deaths in this age group. The rate can increase because the number of suicides increases, or because the number of deaths from other causes decreases, or both. If we are interested in the propensity of this age group to commit suicide, a better base would be the number of persons between ages 15 and 25.

Keep in mind that very small numbers and very large numbers are difficult to read and interpret. Consequently, most rates are adjusted so that the values fall in the range from 0 to 100 or multiples of 100. For example, rather than calculate the proportion of the population that was murdered (which would give extremely small numbers), we use the number of murders per 100,000 population. This is just the proportion murdered multiplied by 100,000. On the other hand, reporting the number of males per 100,000 population would create undesirably large numbers, so the percentage male—the proportion multiplied by 100—is usually used.

Remember that multiplying (or dividing) a rate by a constant value has no effect on the results.

Let's look at how using different bases in constructing rates can affect the results. In the USA data file, variable 27) FS$/PER uses the amount spent on the food stamp program as the numerator and the number of recipients of food stamps as the denominator for each state—so it is the *average benefit per recipient.* The variable 28) FS$/CAP uses the same numerator but uses the population of the state as the base, or denominator—so it is the *average benefit cost per capita.* Let's see if this difference in bases has any effect on relationships with other variables.

➤ *Data File:* **USA**
➤ *Task:* **Scatterplot**
➤ *Dependent Variable:* **28) FS$/CAP**
➤ *Independent Variable:* **32) POV LINE**
➤ *View:* **Reg. Line**

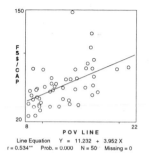

When we look at the scatterplot between the average benefit *cost per capita* and the percentage of people below the poverty line, we see that the correlation is .534 and that it is significant.

Now create a scatterplot between 27) FS$/PER (the average benefit *per recipient*) and 32) POV LINE.

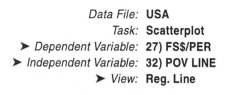

Data File: **USA**
Task: **Scatterplot**
➤ *Dependent Variable:* **27) FS$/PER**
➤ *Independent Variable:* **32) POV LINE**
➤ *View:* **Reg. Line**

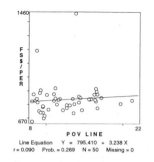

The correlation is only .090 and is not significant. Notice that there are two cases that are at a distance from the others—one is at the top middle and the other one is close to the vertical line about two-thirds of the way up the vertical line. Removing one or both of these cases might have a substantial impact on the scatterplot.

We can use the outlier task to see if there is any single case that is making a substantial difference in the correlation coefficient. Click the [Outlier] option to see the effect of this outlier. The case at the top now has a red square around it. This case has been identified as the case that would have the greatest effect on the correlation coefficient if it were removed. We can see on the left side of the scatter-plot that this case is Alaska and that removal of this case would change the correlation coefficient from 0.090 to 0.183. The correlation would still not be significant. Therefore, if all the other cases seemed to be nonoutliers, then we would have no reason to remove this case; it really wouldn't change the nature of the relationship. However, in this particular situation, we can see that there is another case that is off by itself—the case at the top middle of the screen. So, let's remove Alaska and click the [Outlier] option a second time. Now the outlier is Hawaii. If we removed Hawaii, the correlation would go up to .262 and would be statistically significant. Thus, if we removed the two noncontinental states (Alaska and Hawaii), there is a weak, statistically significant relationship between average food stamp benefit per recipient and the percentage of the population living below the poverty line. However, this relationship is much weaker than the relationship between average food stamp cost per capita and the percentage of the population living below the poverty line, and we only achieved it by removing two states from the analysis.

Clearly in this example, one would reach quite different conclusions by using the cost per participant than by using the cost per capita. Poorer states spend more on food stamps per capita, probably reflecting that a greater percentage of the population receives food stamps. However, the poverty of the state does not appear to affect the amount the average recipient receives.

The impact of Hawaii and Alaska on this correlation coefficient demonstrates another frequent problem in comparative research. Because of the relatively small number of cases, a single case may have a sizable impact on the results. The wise comparative researcher will learn something about the units being used before beginning the actual analysis. In our example, if we knew that food costs for the 48 contiguous states had little variation, and that food in Alaska and Hawaii was much more expensive, we would be better able to interpret the results. We could speculate that Alaska and Hawaii spend more on food stamps because food is much more expensive in those two states. Similarly, if you are using a data set of nations, you should be aware that Singapore and Hong Kong are frequently included as cases in such data sets. However, these are "city-nations"—the boundaries of the city and the boundaries of the nation are the same. Since these cases have virtually no rural or farm areas, they are going to differ in many ways from other nations. Such deviant cases can have sizable effects on the results of an analysis.

With MicroCase, you can determine that an outlier is having a significant impact, but such a determination will not explain why the case is so deviant. In our example, we don't know for sure why Hawaii and Alaska are so different from the other cases and, consequently, we have no justification for removing them from the analysis.

If you have no idea why a particular case would be an outlier, you should double-check the data for that case. First, check the data values for the case against the original source. Perhaps the data were not entered correctly. Even if the data match the source, perhaps the source is incorrect—data are often misprinted or reported incorrectly. Try to find a second source for the data. If the two sources have decidedly different values, you must then determine which is the better source.

In some situations, you will know exactly why a case is an outlier. The results with that case removed may better reflect the overall relationship. You are then justified in removing that case from the analysis. Be sure to report any cases removed from analysis. In some situations, you may want to report the results both with and without the suspect case.

Removing a case from the analysis essentially assigns the missing data value to all variables for that case. This leads us to a third major difficulty in comparative research. Even without removing cases, comparative data sets frequently have a great deal of missing data—even worse, the cases with missing data almost never represent a random selection. For example, prior to the collapse of the Soviet bloc, data were much more difficult to find for Eastern European countries than for the other European countries. Similarly, data for less developed countries are more likely to be incomplete than are data for more developed countries. In the written exercises, you'll get a chance to see how missing data can influence the results.

By this point it should be pretty clear to you why rates, rather than raw numbers, are used in comparative research. In fact, this is a topic that was discussed back in Exercise 3—remember when you compared the *Playboy* circulation rates (circulation per 100,000 population) with the actual circulation rates for the 50 states? Let's take a quick look at this again.

Data File:	**USA**
➤ *Task:*	**Mapping**
➤ *Variable 1:*	**114) PLAYBOY**
➤ *Variable 2:*	**134) PLAYBOY#**
➤ *Views:*	**Map**

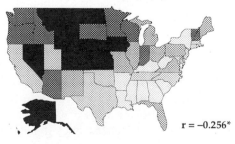

PLAYBOY -- 1996: PLAYBOY CIRCULATION PER 100,000 (ABC)

r = −0.256*

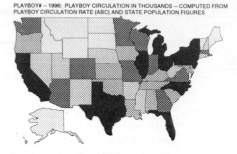

PLAYBOY# -- 1996: PLAYBOY CIRCULATION IN THOUSANDS -- COMPUTED FROM
PLAYBOY CIRCULATION RATE (ABC) AND STATE POPULATION FIGURES

The top map shows the actual circulation of *Playboy* magazine for each state, whereas the bottom map shows the circulation per 100,000 population in each state. If we did not convert state-level variables to rates, nearly all variables we map would look the same because states with the largest populations will generally have more of everything—California has more hockey rinks than North Dakota simply because there are more people in California.

Most aggregate data are not published in the form of rates, so it is up to a researcher to turn variables into rates. Calculating a rate for each case in a data set would be extremely time-consuming if it had to be done manually. But most statistical analysis programs have a recode capability that allows you to compute rates electronically. Student MicroCase has an especially powerful recode capability for creating rates. Let's see how it works.

Go to the **FILE & DATA MENU** and select the RECODE VARIABLES task. Then select the [Rate] option. The next screen prompts you to select the variable you want to turn into a rate. Select 134) PLAYBOY# as your numerator, but pause a moment to look at the variable description for PLAYBOY#. Notice that the coding for this variable is in 1,000s. That means that a state coded as 50 really represents 50,000 copies of the magazine that are sold (you need to move the decimal to the right three places). Hence, in order to properly calculate our rate, we need to indicate that the numerator is in 1,000s. So change the radio button that appears below the numerator box to "in 1,000s." Then click [Next] to continue.

You are now prompted to select your denominator. Since we want population as our denominator, select 2) POP 98 as your denominator variable. Notice again that the description for the denominator indicates that the values are listed in thousands. So we again need to change the setting to "in 1,000s." Once you have done this, click [Next] to continue.

The next step is the trickiest because there are no steadfast rules. On this screen you have the ability to adjust the rate by multiplying either the numerator or denominator by a factor of 10. (Remember, if you multiply both the numerator and denominator by the same number, the ratio or rate remains the same.) You may wonder why it is necessary to adjust the rate at all. The reason for this is to make the results easier to interpret. For example, look at the first column in the grid at the bottom of the screen. This column, which appears in red, shows the results of your current recode: the number of *Playboy* magazines sold per person. Note, however, that these numbers are rather small and that there is not much difference between them—partially because of rounding. Thus, it would be better to convert these numbers into something larger so that we can better see the variation among the states.

As noted, the current multiplier for these results is 1 (which has no effect on the results). Click on the right "spinner" arrow once to increase the multiplier rate by 10. Note that this updates the rest of the information on the screen, including the first column of the grid at the bottom of the screen. Click the right multiplier button again to change the multiplier to 100. Since we have essentially moved the decimal two places to the left by multiplying by 100, the rate is also a percentage. These rates are still going to be hard to interpret because the percentage of people who read *Playboy* ranges from 0.73 percent to 1.95 percent for all the states. So click the right multiplier arrow one more time to make the multiplier rate 1,000. Look again at the results in the first column of the grid. These values seem a little more useful. As you can see, there are approximately 8.96 *Playboy* magazines sold per 1,000 population in Alabama, whereas there are approximately 19.54 sold per 1,000 population in Alaska.

We could keep increasing the multiplier of our rate to 10,000 or 100,000 or more, but let's leave it as a circulation per thousand. Click [Next] to continue. You are now prompted to enter the name of the new variable. Type in the name **PLAYBOY/K** and modify the variable description to your liking. Click [Next] and the following screen indicates that the variable will be added to the end of your data file. Click [Finish] and then [OK] to return to the main menu.

Let's take a look at our handiwork.

Data File: **USA**
Task: **Mapping**
➤ Variable 1: **PLAYBOY/K**
➤ View: **Map**

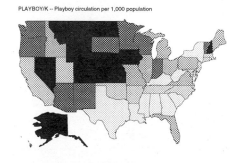

PLAYBOY/K -- Playboy circulation per 1,000 population

It looks just like the map of 114) PLAYBOY we saw earlier, which is based on *Playboy* circulation per 100,000. (You should already know why the maps look identical, even though one is based on the circulation per 1,000 population and the other is based on the circulation per 100,000 population.)

Your turn.

NAME: ____________________

COURSE: ____________________

DATE: ____________________

EXERCISE

8

WORKSHEET

1. You want to compare personal property taxes in different counties of a state. Which would be the best base to compute a rate: the total population of a county, the population over age 18, or the total number required to pay personal property taxes? (Circle one.)

Total population

Population over age 18

Total number required to pay personal property taxes

Explain your answer.

2. You are interested in voter turnout by state in the last presidential election. Obviously, the numerator for the rate would be the number of persons in each state who voted in that election. List three possible bases, or denominators, that you could use for creating a rate.

a. ____________________

b. ____________________

c. ____________________

Circle the letter of the rate you think would be best, and explain why you think it would be the best.

3. Let's examine a variable from the USA data file.

> *Data File:* **USA**
> *Task:* **Mapping**

a. What is the description for 139) FEM-LEGIS?

_____

_____

b. According to the description, what numerator was used in calculating this rate?

_____

c. According to the description, what base, or denominator, was used in calculating this rate?

_____

d. Why wasn't the population of the state used as the base?

4. Traffic statistics use different bases. For example, traffic fatalities can be reported as number of fatalities per 1,000 population, number of fatalities per 1,000 licensed drivers, number of fatalities per 1,000 registered vehicles, and so on. In the USA data set, you have two measures of the amount of driving in each state: 75) MILES/DRV and 76) MILES/VHCL.

Look at the description of each of these variables and describe the difference in how they are calculated.

5. Let's examine several scatterplots in the USA data file.

a. Create the following scatterplot and answer the items that follow.

> Data File: **USA**
> ➤ Task: **Scatterplot**
> ➤ Dependent Variable: **94) PICKUPS**
> ➤ Independent Variable: **75) MILES/DRV**

r = _____

Prob. = _____

Click on [Outlier] to identify the case that has the greatest impact on the correlation coefficient. If this case were removed, what would the results be?

r = _____

Prob. = _____

Should you consider removing the outlier? (Circle one.)   Yes   No

b. Create the scatterplot below and answer the items that follow.

> Data File: **USA**
> Task: **Scatterplot**
> Dependent Variable: **94) PICKUPS**
> ➤ Independent Variable: **76) MILES/VHCL**

r = _____

Prob. = _____

Click on [Outlier] to identify the case that has the greatest impact on the correlation coefficient. If this case were removed, what would the results be?

r = _____

Prob. = _____

Should you consider removing the outlier? (Circle one.)   Yes   No

c. Why does the relationship with 94) PICKUPS differ depending on which measure is used? (Comparing the maps of these variables might help you answer this question.)

6. a. Look at the relationship between 21) #SOC.SEC and 44) ABORTIONS#.

>   Data File: **USA**
>   Task: **Scatterplot**
> ➤ Dependent Variable: **44) ABORTIONS#**
> ➤ Independent Variable: **21) #SOC.SEC**

r = _____

Prob. = _____

   b. Does it make any theoretical sense to expect a relationship between abortions and Social Security recipients? Why is the relationship between these two variables so extremely high? [Hint: Look carefully at the variable definitions.]

   c. Now look at the three abortion rates: 45) ABORTRATE1, 46) ABORTRATE2, and 47) ABORTRATE3. Which one of these do you think is probably the best measure to use to measure the abortion rate for a state? (Circle one.)

ABORTRATE1

ABORTRATE2

ABORTRATE3

   Explain your answer.

   d. Do the following analysis using the abortion rate that you selected as the best.

>   Data File: **USA**
>   Task: **Scatterplot**
> ➤ Dependent Variable: **the rate you selected as the best**
> ➤ Independent Variable: **23) %SOCSEC**

r = _____

Prob. = _____

Is there a statistically significant relationship between
these two variables? (Circle one.)   Yes   No

Why were the results in this analysis so different from the results in the
previous analysis of 44) ABORTIONS# and 21) #SOC.SEC?

7. a. Let's now use mapping to examine influences on the percentage of the
       states' voting age population that voted in the 1996 presidential election.
       First, obtain a mapping for variable 73) %VOTED 96.

> Data File: **USA**
> ➤ Task: **Mapping**
> ➤ Variable 1: **73) % VOTED 96**
> ➤ View: **Map**

Which part of the country has the higher voter
turnout? (Circle one.)

States in the more northern part of the country

States in the more southern part of the country

    b. What influence does urbanism have on voter turnout?
       Would you hypothesize a positive or negative relation-
       ship between degree of urbanism of states and their
       level of voter turnout? (Circle one.)   Positive   Negative

    c. Obtain the following analysis.

> Data File: **USA**
> Task: **Mapping**
> Variable 1:   **73) % VOTED 96**
> ➤ Variable 2: **6) %URBAN**
> ➤ View: **Map**

What is the correlation for this relationship?   r = _____

Is this relationship between urbanism and voter turnout
statistically significant? (Circle one.)   Yes   No

Is the relationship between urbanism and voter turnout
positive or negative? (Circle one.)   Positive   Negative

Do these results support your hypothesis? (Circle one.)   Yes   No

d. Now let's examine the link between the unemployment rate
in states and their voter turnout in the 1996 presidential
election. Would you hypothesize a positive or negative
relationship between these two variables? (Circle one.)   Yes   No

e. Obtain the following analysis.

> Data File: **USA**
> Task: **Mapping**
> Variable 1: **73) % VOTED 96**
> ➤ Variable 2: **95) UNEMPLOY**
> ➤ View: **Map**

What is the correlation for this relationship?   r = _____

Is this relationship between unemployment rate and
voter turnout statistically significant? (Circle one.)   Yes   No

Is the relationship between unemployment rate and
voter turnout positive or negative? (Circle one.)   Positive   Negative

Do these results support your hypothesis? (Circle one.)   Yes   No

8. a. We will now switch to scatterplot and examine the relationship between the
syphilis rate in states and the percentage of the population that doesn't have
health insurance. Which of the following would you hypothesize? (Circle
the number of your choice.)

   1. Positive relationship: States that have higher percentages of people
   who have no health insurance will have a higher percentage of the
   population with syphilis.

   2. Negative relationship: States that have higher percentages of people
   who have no health insurance will have a lower percentage of the population with syphilis.

   Explain why you expect to find the relationship you hypothesized.

b. Obtain the following analysis and write in the correlation and significance level.

              Data File: **USA**
        ➤ *Task:*   **Scatterplot**
  ➤ *Dependent Variable:*   **8) SYPHILIS**
 ➤ *Independent Variable:*   **14) HEALTH INS**

r = _____

Prob. = _____

Is this relationship statistically significant? (Circle one.)   Yes   No

Is the relationship between syphilis rate and
percentage of the population without health
insurance positive or negative? (Circle one.)   Positive   Negative

Do these results support your hypothesis? (Circle one.)   Yes   No

9.   Sometimes, researchers will include Washington, D.C., in a data set with the 50 states. Let's examine the data set named US–DC.

          ➤ *Data File:*   **US–DC**
         ➤ *Task:*   **Scatterplot**
  ➤ *Dependent Variable:*   **3) %GRAD.DEG**
 ➤ *Independent Variable:*   **2) %BLACK**

a. Based on this scatterplot, fill in the following:

r = _____

Prob. = _____

b. Click on [Outlier] to identify the case that has the greatest impact on the correlation coefficient.

Name of outlier: _____

r = _____

Prob. = _____

c. Should you consider removing the outlier? (Circle one.)   Yes   No

d. What is your conclusion about the relationship between these two variables?

e. Create the scatterplot below and answer the items that follow.

> Data File: **US–DC**
> Task: **Scatterplot**
> ➤ Dependent Variable: **4) ONE P.HH**
> ➤ Independent Variable: **3) %GRAD.DEG**

What is the description of 4) ONE P.HH?

_____

_____

r = _____

Prob. = _____

f. Click on [Outlier] to identify the case that has the greatest impact on the correlation coefficient.

Name of outlier: _____

r = _____

Prob. = _____

Should you consider removing the outlier? (Circle one.)   Yes   No

g. What is your conclusion about the relationship between these two variables?

h. Would you advise researchers to include Washington, D.C., in a data set on the states of the United States? (Circle one.)   Yes   No

Explain your answer below.

10. Let's do another recode. In the USA data file, variable 2) POP 98 is population in thousands and 135) AREA is area in square miles. So, let's compute a population density figure by dividing population in thousands by area in square miles.

> ➤ *Data File:*   **USA**
> ➤ *Task:*   **Recode**
> ➤ *Recode Option:*   **Rate**

For the numerator, select 2) POP 98. Since our population variable is in thousands, click on the radio button that indicates "in 1,000s" and click [Next]. For the denominator, select 135) AREA and click [Next]. This yields a rate that shows the population in thousands per square mile. Convert this to population per square mile by clicking the right "spinner" button three times. Note that the number in the Multiplier column has changed to 1,000.

How many people per square mile are there in each of the following states?

Alaska   _____

Kentucky   _____

New York   _____

Click [Next] and provide a name for this new variable (you are limited to 10 characters). Click [Next], [Finish], and then [OK] to return to the main menu.

> *Data File:*   **USA**
> ➤ *Task:*   **Mapping**
> ➤ *Primary Variable:*   **Select the variable you created**
> ➤ *View:*   **List: Rank**

Print out this table and attach it to this worksheet.

11. Lastly, let's do some elementary time series analysis by examining the relationship between political party identification and gender over time. It has been noted that a small but important gender difference (the "gender gap")

in political party identification has developed whereas it did not exist at an earlier time. To examine this proposition, we will use results from GSS surveys for every four years since 1972 (except that 1993 is used instead of 1992 because there was no GSS in 1992).

Using the 1972 to 1996 GSS files, the party identification variable (Democrat, Independent, or Republican) was cross-tabulated by gender for each of the selected years. Here, we will just present the percentages of males and females identifying as Democrats. (Note: For present purposes, Independents who leaned toward a particular party were grouped with the party toward which they leaned.)

The percentages of males and females identifying with the Democratic Party are given below. The "Gap"column is the percentage of females identifying as Democrats minus the percentage of males identifying as Democrats. An asterisk next to the gap means that the relationship between party identification and gender is statistically significant.

Percent Democrats

Year	Males	Females	Gap
1972	60%	61%	1%
1976	55%	57%	2%
1980	51%	52%	1%
1984	49%	54%	5% *
1988	47%	51%	4%
1993	44%	49%	5% *
1996	40%	52%	12% *

a. First, examine the percentages of males identifying as Democrats from 1972 to 1996. What kind of trend is there here?

b. Next, examine the percentages of females identifying as Democrats from 1972 to 1996. What kind of trend is there here?

c. Now look at the gap between men and women in terms of percentages identifying with the Democratic Party. What kind of trend is there?

9

Field Methods

OVERVIEW

In this exercise, you will learn more about problems and issues in field research. Field research is less structured than other methods, and this has both advantages and disadvantages. You will see that field researchers must be careful not to let this lack of structure undermine their studies.

BEFORE YOU BEGIN

Please make sure you have read Chapter 9 in the textbook and can answer the following review questions (you need not write any answers):

1. When is field research better than other methods of social research?

2. Why is reliability often low in field studies, and how can it be increased?

3. Discuss the two major problems involved in entering the field.

4. What is the difference between structured and unstructured observation?

5. What are informants in field research? What are the advantages and disadvantages of using informants?

6. Describe field notes and the general processes by which they are analyzed.

7. What are the primary ethical concerns in field research?

Field research is considerably less structured than other types of social research. In survey research, we can look at how the sample was selected and at the questions asked. In comparative research, we can determine whether a particular rate appears to be an acceptable measure. But, in reports of field research, we have virtually no means of checking on the accuracy of the observations. This is why most field studies are exploratory, rather than hypothesis testing. In this exercise, we'll see some special problems that are created by this lack of structure.

One of the first problems is that different observers focus on different elements. For example, consider something as uncomplicated as a group of people talking with one another. One researcher might focus on the content of the conversation, while another might be more interested in nonverbal communication. Differences of style between males and females might be the main interest of another. Decoding the influence, or power structure, of the group might be another focus. Obviously, the interest of the researcher will influence what is observed and recorded, as well as what is not observed and not recorded. Of course, the purpose of field research is not only to observe, but also to organize and interpret these observations in a meaningful way.

Unfortunately, many researchers not only have certain interests, but also have developed definite opinions on a topic. For example, the researcher interested in gender styles of interaction already may believe that females will be submissive and males will be dominant. Bias—either intentional or unintentional—is almost certain to creep into this study. This researcher already has a hypothesis and is not using field research to explore gender styles in interaction but rather to document preconceived differences. This is not exploratory research but hypothesis-testing research. In this situation, other methods better suited to hypothesis testing would be more appropriate—content analysis of videotaped conversations might be one such approach.

At the other extreme, researchers who observe without any preconceived structure are bound to be buried in a sea of unrelated details and accomplish nothing. Thus the problem for the field researcher is to focus the study without predetermining its outcome.

Characteristics of the researcher also can have a major effect on the subjects of observation. An individual's actions are frequently influenced by who is watching. All of us engage in private behaviors that we would not want observed. We behave differently in front of our boss than we do in front of a subordinate. Age, gender, attractiveness, and many other characteristics of the observer can influence the behavior of those being observed. A 20-year-old woman and a 50-year-old man observing the same group might reach quite different conclusions since the behavior of the group may be affected by who is watching. In Japan, even the spoken language differs greatly according to the gender of the speaker; that is, men and women have different accents. This led to considerable embarrassment for some World War II American GIs who learned the language from their Japanese girlfriends. This is another reason that having more than one observer is a good idea.

Obviously, the effects of observer characteristics will be even more pronounced when informers are involved. In some societies, age, gender, and marital status determine interaction patterns—some members may not even be permitted to talk with the observer.

Field research also may be applied research. A researcher might observe the operation of a ward in a hospital. Sometimes, such studies are unstructured. For example, the researcher may discover that doctor/nurse conflicts are having a negative impact on patient care. More often, they are studies to *evaluate* an organization or process. The researchers already have a model about what "should" happen. The observations are used to determine the extent to which this "model" is being met. For example, business consultants frequently observe the operation of a company for a period of time in order to recommend changes that might improve the operation of the business. Typically, such researchers already know how "good" companies work—for example, that employees are sharing in the decision making. Their observations are used to determine how the company should be changed to match this preconceived ideal. Such studies may be very useful for the organizations involved, but they are not social science.

1. You want to study the ways in which a local public official tries to win re-election. There are at least two fundamentally different approaches to gaining access. Describe two approaches and the advantages and disadvantages of each approach.

    a. Describe the first approach.

        What are the advantages of the first approach?

        What are the disadvantages of the first approach?

    b. Describe the second approach.

        What are the advantages of the second approach?

        What are the disadvantages of the second approach?

2. Look at this photograph of new recruits to the Marine Corps to answer the questions that follow.

a. How would a field researcher obtain access to this setting?

b. Boot camp is eight weeks long. What problems does this pose for the researcher?

c. What characteristics of the researcher might affect his or her success in completing a study in this setting?

d. Discuss the relative value of a new recruit versus that of an instructor as an informer.

e. List three different topics that a field researcher might study in Marine Corps basic training.

   1.

   2.

   3.

f. What problems exist for taking field notes in a study of basic training?

g. If the recruits were female, how would this affect the research project?

3. A field researcher who is observing police behavior remarks that she is sure the police officers are acting normally around her because they frequently use foul language and they often criticize their supervisors. Do you agree with her conclusion? Why or why not?

4. a. A team of researchers is interested in studying the role of house-husband/stay-at-home dad. They are especially interested in how this role differs from that of the housewife/stay-at-home mother. How would you design a field study to explore this topic?

    b. List two problems the field researchers are likely to encounter and how you would solve them.

        1.

        2.

5. Consider the following potential "informants" in a study of police behavior in a small city. Discuss the pluses and minuses of trusting the information of each informant listed.

   a. The mayor of the city

   b. The chief of police in this city

   c. The prosecutor in this city

   d. A defense lawyer

   e. A prisoner in the city jail

6. A researcher pretends to be a supporter in order to infiltrate an antigovernment organization that sometimes commits illegal acts. What ethical problems are most relevant here?

7. A social scientist who is a member of an organization decides to do a field study of this organization and receives funding from the organization for the study. Describe the problems in this situation.

8. A social scientist who has very strong views on the abortion controversy decides to do a field study of an organization that works for the other side on this issue. Describe the problems in this situation.

Experimental Methods

OVERVIEW

In this exercise, you will learn more about the necessary conditions for an experiment and about the power of the experiment in eliminating sources of spuriousness in relationships. You also will gain experience in designing experimental research and drawing the proper conclusions from the results.

BEFORE YOU BEGIN

Please make sure you have read Chapter 10 in the textbook and can answer the following review questions (you need not write any answers):

1. What are the two essential characteristics of true experiments?

2. What advantages do laboratory experiments have over field experiments? What advantages do field experiments have over laboratory experiments?

3. What is the difference between an experimental group and a control group?

4. What is a double-blind experiment, and what use does it serve?

5. How can experiments be used in surveys?

6. What two primary sources of bias in experiments can reduce reliability?

7. Why is it important to make sure in an experiment that the independent variable really varied in the way it was supposed to vary?

8. Describe how each of the following factors can affect the results of quasi-experiments: history, maturation, testing effects, instrument effects, regression to the mean, and selection bias.

9. What is subject "mortality," and how can it affect experimental results?

10. What are the primary ethical concerns in experimental research?

The social scientists who designed the questionnaire for the 1998 General
Social Survey were interested in how question wordings on government
spending might affect the responses—would relatively minor changes in the
phrasing of questions have a significant impact on the distribution of responses?
They recognized that an experiment would be the best method for testing for such
an effect. Half of the survey respondents, randomly selected, would be given one
question wording, and the other half would be given the second question wording. The responses of the two groups could then be compared. The independent
variable would be the question wording—there are two different treatment conditions represented by the two different wordings. The dependent variable would
be the response to the question. By comparing the responses of subjects in the two
treatment conditions, researchers could tell whether question wording had an effect.

Let's analyze the results of this experiment. Because the GSS is a survey, the
GSS data file is organized in a manner appropriate for survey analysis.
Consequently, we have created a data set named EXPER that is better organized
for experimental analysis.

> *Data File:* **EXPER**
> > *Task:* **Cross-tabulation**

Let's examine the variable list for this file. The value of the first variable is
the experimental condition for that subject—were the respondents asked the first
wording of the question or the second wording? The second variable (WELFARE)
is the response to the first question. The first treatment used the following wording: Are we spending too much, too little, or about the right amount on *welfare*?
The second treatment had different wording: Are we spending too much, too
little, or about the right amount on *assistance to the poor*? We want to compare the
responses of those in treatment 1 with the responses of those in treatment 2. You
can see that there are a series of additional questions about spending in which the
wording differed slightly between the two treatment conditions. So, in fact, we
have a series of experiments.

Because the cases were *randomly assigned* to conditions, other characteristics
of the cases will not be correlated with the independent variable (i.e., the question
wording). We can see this by looking at the cross-tabulation of the treatment variable with these other characteristics.

> *Data File:* **EXPER**
> *Task:* **Cross-tabulation**
> > *Row Variable:* **13) SEX**
> > *Column Variable:* **1) TREATMENT**
> > > *View:* **Tables**
> > > *Display:* **Column %**

SEX   by   TREATMENT
Cramer's V: 0.003

		TREATMENT			
		One	Two	Missing	TOTAL
SEX	Male	575	611	46	1186
		43.7%	44.0%		43.8%
	Female	742	779	79	1521
		56.3%	56.0%		56.2%
	TOTAL	1317	1390	125	2707
		100.0%	100.0%		

We can see that the distribution of sex is almost identical in the two conditions—43.7 percent of the subjects in the first condition and 44.0 percent of the subjects in the second condition were male. Other characteristics of the subjects would also be about the same across the two conditions. The fact that the independent variable is uncorrelated with other characteristics of the subjects is the strength of experimental research. Remember, for a third variable to make a relationship spurious, that variable must be correlated with both the independent and dependent variables. So in experiments we do not have to worry about possible sources of spuriousness. We only need to examine the relationship between the independent and dependent variables to support or reject our research hypothesis.

Data File: **EXPER**	WELFARE by TREATMENT
Task: **Cross-tabulation**	Cramer's V: 0.492 **
➤ *Row Variable:* **2) WELFARE**	
➤ *Column Variable:* **1) TREATMENT**	
➤ *View:* **Tables**	
➤ *Display:* **Column %**	

WELFARE by TREATMENT
Cramer's V: 0.492 **

WELFARE	TREATMENT			
	One	Two	Missing	TOTAL
Too little	221	866	0	1087
	16.8%	62.3%		40.2%
Right	498	364	0	862
	37.8%	26.2%		31.8%
Too much	598	160	0	758
	45.4%	11.5%		28.0%
Missing	0	0	125	125
TOTAL	1317	1390	125	2707
	100.0%	100.0%		

We can see that 16.8 percent of those shown the first wording (i.e., *spending on welfare*) thought too little was being spent, while 62.3 percent of those in the second treatment condition (i.e., *assistance to the poor*) felt this way. Let's look at the statistical summary.

Data File: **EXPER**
Task: **Cross-tabulation**
Row Variable: **2) WELFARE**
Column Variable: **1) TREATMENT**
➤ *View:* **Statistics (Summary)**

WELFARE by TREATMENT

Nominal Statistics

Chi-Square: 855.159 (DF = 2; Prob. = 0.000)

V:	0.492	C:	0.441		
Lambda: (DV=1)	0.434	Lambda: (DV=2)	0.233	Lambda:	0.323

Ordinal Statistics

Gamma:	-0.713	Tau-b:	-0.462	Tau-c:	-0.530
s.error	0.023	s.error	0.015	s.error	0.017
Dyx:	-0.531	Dxy:	-0.402		
s.error	0.017	s.error	0.013		
Prob. =	0.000				

We can see that the results are statistically significant. So the difference in wording did have an effect on the subject's view toward government spending.

If we can assume that the dependent variable is measured at the interval or ratio level, there is another technique that can be used in experimental analysis. Analysis of variance is similar to regression analysis except the independent variable is nominal, or categorical. To demonstrate this technique, we'll assume that the dependent variables are close enough to interval to justify its use.

Data File: **EXPER**
➤ *Task:* **ANOVA**
➤ *Dependent Variable:* **2) WELFARE**
➤ *Independent Variable:* **1) TREATMENT**
➤ *View:* **Graph**

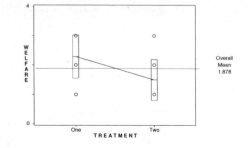

This is called a *box-and-whisker diagram*. It is just a scatterplot between the two variables with some additional information. The independent variable, treatment condition, is shown across the bottom. Those in the first treatment condition are shown at the left, and those in the second treatment condition are shown at the right. The vertical axis represents the dependent variable. In this example, there are only three possible values for this variable, so there are many cases at each of the three levels but only one dot is visible. So far, this is just the same as a scatterplot.

Since the independent variable in analysis of variance is categorical, we can provide additional information about the dependent variable within each of these categories. We can calculate the mean and variation of the dependent variable within each category of the independent variable. The vertical box shown at the left represents the variation in the dependent variable for the first condition. The horizontal line (the whisker) in the middle of this box represents the mean of the dependent variable. The box at the right shows the same information for the second condition. A line connects the means of the two conditions. In this example, there is a sizable difference in the means.

Let's look at the analysis of variance (ANOVA) results.

Data File: **EXPER**
Task: **ANOVA**
Dependent Variable: **2) WELFARE**
Independent Variable: **1) TREATMENT**
➤ *View:* **ANOVA**

Analysis Of Variance
Dependent Variable: WELFARE
Independent Variable: TREATMENT
N: 2707   Missing: 125

ETA Square = 0.236

TEST FOR NON-LINEARITY:
Not appropriate when only 2 categories for Independent Variable.

Source	Sum of Squares	DF	Mean Square	F	Prob.
Between	426.520	1	426.520	836.955	0.000
Within	1378.494	2705	0.510		
TOTAL	1805.014	2706			

This is similar to the analysis of variance table shown for regression. For our needs, we can ignore all the information except the significance level (Prob. = 0.000) presented next to the F value. The eta-squared value (0.236) is a measure of association and tells us the strength of the relationship between the two variables.[1] In this example, this relationship is highly significant. Now select the [Means] option.

Data File: **EXPER**
Task: **ANOVA**
Dependent Variable: **2) WELFARE**
Independent Variable: **1) TREATMENT**
➤ View: **Means**

Means, Standard Deviations and Number of Cases of Dependent Var:
WELFARE
by Categories of Independent Var: TREATMENT
Difference of means across groups is statistically significant (Prob. = 0.000)

	N	Mean	Std.Dev.
One	1317	2.286	0.735
Two	1390	1.492	0.693

This is the mean of the dependent variable for each condition. For the 1,317 cases in the first treatment condition, the mean was 2.286—just slightly under the midpoint for the categories "right" and "too much." For the 1,390 cases in the second condition, the mean was 1.492—or just slightly above the midpoint for the categories "too little" and "right."

Let's now look at the wording of two questions relating to spending on the space program. Again, half the respondents were asked one version of the question (government spending on the *space exploration program*), while the second half were asked another version (government spending on *space exploration*).

Data File: **EXPER**
Task: **ANOVA**
➤ Dependent Variable: **3) SPACE PRG**
➤ Independent Variable: **1) TREATMENT**
➤ View: **Graph**

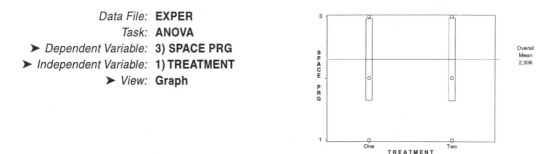

In this example, the line that connects the means is horizontal. There is virtually no difference between the conditions. If you look at the analysis of variance table, you see that the results are not significant and eta-squared is zero.

Here is another set of questions relating to government spending on health. (Examine the variable description for 8) DRUGS to see the difference in question wording.)

---

[1] Eta and r, the Pearson correlation coefficient, are based on the same mathematical ideas except that r assumes that the relationship will be linear and eta has no restrictions on the nature of the relationship.

Data File: **EXPER**
Task: **ANOVA**
➤ Dependent Variable: **8) DRUGS**
➤ Independent Variable: **1) TREATMENT**
➤ View: **Graph**

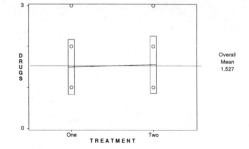

Notice that the line connecting means is almost horizontal. Let's look at the actual means.

Data File: **EXPER**
Task: **ANOVA**
Dependent Variable: **8) DRUGS**
Independent Variable: **1) TREATMENT**
➤ View: **Means**

Means, Standard Deviations and Number of Cases of Dependent Var: DRUGS

by Categories of Independent Var: TREATMENT
Difference of means across groups is statistically significant (Prob. = 0.037)

	N	Mean	Std.Dev.
One	1267	1.499	0.673
Two	1302	1.555	0.700

Treatment 1 with 1,267 cases has a mean of 1.499, while treatment 2 with 1,302 cases has a mean of 1.555. Not much difference. Now let's look at the analysis of variance table.

Data File: **EXPER**
Task: **ANOVA**
Dependent Variable: **8) DRUGS**
Independent Variable: **1) TREATMENT**
➤ View: **ANOVA**

Analysis Of Variance
Dependent Variable: DRUGS
Independent Variable: TREATMENT
N: 2569   Missing: 263

ETA Square = 0.002

TEST FOR NON-LINEARITY:
Not appropriate when only 2 categories for Independent Variable.

Source	Sum of Squares	DF	Mean Square	F	Prob.
Between	2.049	1	2.049	4.345	0.037
Within	1210.267	2567	0.471		
TOTAL	1212.315	2568			

Eta-squared is .002 and is significant at the .037 level. How can such a weak relationship be significant?

Remember that statistical significance tells us only that it is likely that a relationship exists in the population. Two elements affect this level: the strength of the relationship in the sample and the size of the sample. In this study, the sample size is extremely large so that even a small relationship is likely to be significant. Again this demonstrates why we should look at both the significance level and the strength of the relationship. The eta-squared of .002 tells us that, while a relationship may exist in the population, it is a very weak relationship.

We can see that the analysis of experiments is much simpler than the analysis of surveys because we need not worry about sources of spuriousness.

What do these results tell us? Simply that relatively minor changes in wording can have a significant effect on responses. (We examined this issue in a slightly different way in Exercise 3.) Suppose that instead of doing an experiment, researchers had used one set of question wordings in the GSS one year and the other set of wordings the next year. If we observed a difference in the responses between the two years, we could not be sure that it was caused by the change in question wordings. It is possible that the attitudes in the population actually changed during the year. By using an experiment, we have eliminated any such alternative explanations and can be sure that the question wordings themselves are causing the differences.

This experiment is unusual in that a probability sample of a particular population provided the subjects for the experiment. This means that the results of this study can be generalized to the relevant population. However, this isn't really very important because our interest wasn't in this specific population but in the more general issue of the importance of question wordings. The use of a probability sample also means that it is possible to examine the effect of other variables, such as education, income, and race, on attitudes toward spending. Such analysis, of course, would be regular survey analysis and you would have to consider possible sources of spuriousness.

Your turn.

1. Suppose that a particular state is trying to reduce the percentage of students that drop out before completing high school and that the state legislature has allocated funds to study the situation. Methods have already been devised to identify those students who are at risk for dropping out, and a list of such students has been developed. The head of the state school system hired you to do an experimental study to determine whether providing a limited amount of tutoring (one hour per week) would keep these at-risk students from dropping out of high school. You have the funds and the authority to carry out an experiment to determine whether this limited amount of tutoring will help keep at-risk students in school.

   a. Describe how you would do this experiment.

   b. Draw a diagram (similar to the relevant diagram in the textbook) to show the structure of your experiment.

2. The text describes an experiment in which students are randomly assigned to an experimental group that, over a period of six months, watches a series of movies designed to reduce racial prejudice or to a control group that does not watch the movies.

   a. Suppose a large number of subjects dropped out of the experimental condition—they stopped staying after school to watch the movies. What problem would this introduce in reaching a conclusion about the effect of the movies?

   b. How could the design of the experiment be changed to minimize this problem?

   c. Suppose that on the posttest there was no difference in racial prejudice between the two groups—individuals in both groups were extremely low in prejudice. In addition, there was no change between pretest and posttest for individuals in either group. Would this indicate that the movies had no effect on prejudice? Why or why not?

   d. How could this study be modified to provide a better test of the effectiveness of the movies?

e. Suppose that on the posttest there was no difference in racial prejudice between the two groups—individuals in both groups were low in prejudice. However, in both groups, there was a big decrease in racial prejudice between the pretest and the posttest for each individual. Analyzing the friendship structure and interpersonal contact among all subjects, the researchers find that subjects in the control group had many friends and a great deal of interpersonal contact with subjects in the experimental group. How might this fact have affected the results of the study?

f. How could the researchers modify this study to eliminate this problem?

3. Two social scientists were interested in studying the effect of sex education on sexual behavior. They selected a school in which some of the students would be participating in a sex education study during the year while others would be taking a computer literacy course. (Parents selected one or the other of these courses for their children.) The researchers administered a questionnaire on sexual behavior at the end of the year and found that students who had taken the sex education sequence were much more likely to be sexually active than were the other students. They concluded that sex education increases the sexual activity of preteens.

a. Given the design of this study, can we have any confidence in the conclusion? (Circle one.)   Yes   No

When the researchers were told that this was not an experiment, they responded, "That's OK. It's a quasi-experiment."

b. Describe why this study is not a true experiment.

c. Provide another explanation for why the researchers found the two groups to differ in sexual activity.

d. Assume that the researchers' hypothesis was that taking a sex education class would increase the sexual activity of the participants. How could they have designed a true experiment to test this hypothesis?

e. What ethical problems would the researchers encounter in this study, and how could they solve them?

4. Does taking a college course on American politics increase the political knowledge of students? Let's examine two approaches to answering this research question.

a. At the end of the first term of the school year, a researcher measures political knowledge among two samples of first-year students: a sample of those who took the American politics course and a sample of those who did not take the course. Students who took the course have substantially higher political knowledge than those who did not take the course. The researcher concluded that taking the course on American politics does increase the political knowledge of students. What flaws are there in this research design?

b. A second researcher uses a different approach. At the beginning of the term, the researcher measures political knowledge among all students who are taking the American politics course. At the end of the course, the researcher again measures political knowledge among all students who completed the course. At the end of the course, students who completed the course have much higher political knowledge than they did at the beginning of the course. The researcher concludes that taking the American politics course does increase the political knowledge of students. What flaws are there in this research design?

5. Now use the EXPER file to fill in the column percentages in the table below.

> ➤ *Data File:* **EXPER**
> ➤ *Task:* **Cross-tabulation**
> ➤ *Row Variable:* **14) RACE**
> ➤ *Column Variable:* **1) TREATMENT**
> ➤ *View:* **Tables**
> ➤ *Display:* **Column %**

	One	Two
White	_____%	_____%
Black	_____%	_____%
Other	_____%	_____%

V = _____

Prob. = _____

Why is the racial distribution so similar across the two treatment conditions?

6. Variables 4 to 7 and 9 to 12 were also part of the methodological experiment. Let's return to the ANOVA task.

    a. Select one variable from among variables 4 to 12 (except variable 8, DRUGS, which you have already used) as the dependent variable, and use 1) TREATMENT as the independent variable.

    Which variable did you select?   _____

    What is the description of this variable?

    _____

    _____

    What is the mean of each group?

    Treatment 1: mean is _____

    Treatment 2: mean is _____

    Is this difference statistically significant? (Circle one.)   Yes   No

    b. Select a second variable from among variables 4 to 12, and conduct the exact same analysis.

    Which variable did you select?   _____

    What is the description of this variable?

    _____

    _____

    What is the mean of each group?

    Treatment 1: mean is _____

    Treatment 2: mean is _____

    Is this difference statistically significant? (Circle one.)   Yes   No

7. In the first class session for a course, the professor announces that each person in the class has been randomly assigned to one of two groups, A or B. Group A will use the regular textbook for the course, but group B will use a computerized tutorial—a tutorial that the professor has developed and believes will greatly increase student learning while also making it much more enjoyable. The two groups will otherwise receive the same treatment (same lectures, exams, etc.). The professor announces that the purpose of this experiment is to determine whether the computerized tutorial increases learning. At the end of the course, the professor will compare the exam scores of the two groups to determine whether the computerized tutorial was more effective than the textbook.

a. Describe how subject bias (demand characteristics) could be a problem here.

b. Describe how experimenter bias could be a problem here.

c. What ethical issue might be raised by some of the students in the course? How might this problem be resolved?

Content Analysis

OVERVIEW

In this exercise, you will learn more about content analysis by doing a small content analysis of newspaper personal ads in which people are trying to meet people of the opposite sex. You will code some of the variables and then test several hypotheses based on the data file.

BEFORE YOU BEGIN

Please make sure you have read Chapter 11 in the textbook and can answer the following review questions (you need not write any answers):

1. What is content analysis, and what kinds of materials can be used in such analysis?

2. What is the difference between manifest content and latent content?

3. Describe the two primary factors that determine reliability in content analysis.

4. How do deposit bias and survival bias affect the validity of content analysis?

5. What does it mean to say that content analysis is a form of unobtrusive measurement?

The unit of analysis in content analysis is not an individual, nor even a collection of individuals, but rather a cultural artifact. However, the basic steps of the research process remain the same. After identifying the topic of interest and developing the research hypotheses, the researcher selects an appropriate sample of cases (artifacts), develops measures of the concepts (the coding system), collects the data (codes the cases), and tests the hypotheses.

In this exercise, you will apply content analysis to personal ads in which people are seeking to meet persons of the opposite sex. These ads were selected from one issue of a daily newspaper in the Pacific Northwest. In this particular newspaper, the newspaper ads are free. Each advertiser records a voice greeting that plays when an interested person calls the published number. Those who respond to the ads pay $1.99 per minute to hear the message at the 1-900 number and to leave a message for the advertiser.

Since there were an equal number of male and female ads, a stratified random sampling design was used. The ads were first divided into those placed by men and those placed by women. Then 50 ads were selected randomly from each gender group. The ads included in the sample are reprinted at the end of this discussion—the first 50 are written by males and the second 50 are written by females.

Most of the ads provide information about the ad writer and describe characteristics of the person he or she is seeking. The content of these ads can be coded easily. In fact, if you look at the top of page, you'll see that some elements already have been coded by the advertiser. For example, W means "white," C means "Christian," LTR means "long-term relationship," and so on. Looking over the ads, we decided upon the following variables and categories:

GENDER
Gender of advertiser:
>   1—male, 2—female

AGE-AD
Age of advertiser:
>   1—under 40, 2—40 & over, 3—not stated
>   If range given, code youngest end.

AGE-DES
Age desired:
>   1—under 40, 2—40 & over, 3—not stated
>   If range given, code youngest end.

LTR
Long-term relationship (LTR):
>   1—LTR stated, 2—not clear, 3—no long term
>   If indicate "friends first," code as 2.

SEXUAL IMP
Sexual implication:
>   1—explicit, 2—implicit, 3—none
>   If indicate "friends first," code as 3.

APPEAR-AD
Physical description of advertiser:
>   1—none provided, 2—minimal description provided (e.g., height/ weight proportionate or height), 3—stresses attractiveness (e.g., cute, beautiful, handsome, good looking)

APPEAR-DES
Physical description desired:
>   1—none provided, 2—minimal description provided (e.g., height/ weight proportionate or height), 3—stresses attractiveness (e.g., cute, beautiful, handsome, good looking)

SECUR
Financial security desired:
　　　　1—stated, 2—none indicated
　　　　If seeking professional, code as 1.
ACTIVE
Stresses active lifestyle:
　　　　1—yes, 2—no

The data file for this project is named CONTENT. Open this file and select the LIST DATA task.

➤ *Data File:* **CONTENT**
　➤ *Task:* **List Data**

Go to the FILE & DATA MENU and select LIST DATA. Then click [OK] to accept the settings on the pop-up window.

The program will now list the data for all variables and all cases. Each case is one of the ads. The order of the cases matches the order in the list of ads. The variables have been defined and data have been entered for most of the variables.

Let's look at the coding of the first case:

1. CHRISTIAN AND KIND
SWM, 23, Christian values, 6'1", physically fit, attractive, enjoys summer activities, beaches, conversation. Would like to meet a Christian female for friendship, possibly more.

We now can examine the codes for each variable. (Use the scroll bar to move to the variables not currently showing.) This case has been coded 1 for male for the first variable (GENDER); variable 2 (AGE-AD) is coded as 1, which represents the under 40 category, because the age of the advertiser is indicated as 23; the desired age is not stated for the first case, so variable 3 (AGE-DES) is coded as 3. Notice how this coding is matching the row for the first case in the data set. The ad states that the advertiser is looking for "friendship, possibly more." This is coded as a 2 according to the "friendship first" note for variable 4. There are no sexual implications in the ad, so variable 5 (SEXUAL IMP) is coded 3. The advertiser describes himself as "attractive"—this places him in category 3 of variable 6 (APPEAR-AD). He mentions neither the appearance nor the financial security of the person he is seeking, so variable 7 (APPEAR-DES) is coded as 1 and variable 8 (SECUR) is coded as 2. Finally, he is coded as 1 on variable 9 (ACTIVE). If you were to list the data for all cases, you would see that each case has already been coded for these first nine variables.

We're going to test the hypothesis that the sex of the advertiser affects the content of the ads. But first, we need to determine if our coding is reliable enough for analysis. In a previous exercise, you learned about the use of rekey verification to increase the accuracy of data entry from questionnaires. Rekey verification is useful for checking data entry mistakes—the correct value is obvious. In content analysis, the "correct" answer is subject to interpretation. The researcher wants to be able to compare the data from multiple coders and assess the degree of concensus. Data from each of the coders must be available to the researcher, so rekey verification is not an appropriate technique. Instead, each additional coder should enter data into new variables.

Note that no data have been entered for variables 10 to 18. These variables are essentially the same as variables 1 to 9, except that the variable names include an extra character. These variables will be used by a second coder (that being you) to reenter the same classified advertising data. In this way you will be able to compare your coding to that determined (and entered) by the original coder. Again, by comparing the coding of the same data by two different coders, we can assess the reliability and accuracy of our data. In the exercises, you'll be asked to code several cases on some of these variables, so let's quickly see how to enter the data using MicroCase.

Exit the LIST DATA task and return to the main menu. With the CONTENT data file open, select the ENTER DATA task.

> Data File: **CONTENT**
> ➤ Task: **Enter Data**

A pop-up window will appear when you select the ENTER DATA task. Click [OK] to accept the default settings as Grid Entry format. Then click on the cell in the upper left corner to position the cursor for data entry. Make sure that case #1 is showing in the upper left corner of the grid. If it is not, scroll up to the beginning of the grid.

Let's enter the codes (i.e., the numeric categories) for the first two cases for variables 10 through 18. (Use the scroll bar to move to the variables not currently showing.) We just discussed the codes for the first case, so we'll reenter those codes for variables 10 through 18. Here is a summary of how the first case was coded on each of the nine variables:

	Code	Category Label
GENDER1	1	male
AGE-AD1	1	under 40
AGE-DES1	3	not stated
LTR1	2	not clear
SEXUAL IM1	3	none
APPEAR-AD1	3	stresses attractiveness
APPEAR-DE1	1	none provided
SECUR1	2	none indicated
ACTIVE1	1	yes

After you've entered the code for variable 18, the program will go to variable 10 for case 2. The first case has a value of 1 for the variable 10, GENDER1. To enter this value, simply type **1** and press the <Enter> key. The cursor will move to the next variable, AGE-AD1, for the first case. Type **1** and press the <Enter> key. Enter values for the remaining variables in the same manner. If you make a mistake, just click on the appropriate cell and reenter the value. If you accidentally enter a value which is out of the appropriate range, the program will refuse to accept the value. For example, if you try to use the value 3 for the gender of the individual, a message will tell you that this value is out of range—you will then want to enter the correct value. Enter the remaining data for the first case.

Let's take a look at the second case and code it.

2. CHRISTIAN IN SEATTLE
Attractive, Christian SWM, 28, 6', H/W proportionate, active person, financially secure, stable and healthy. Would like to meet Christian female, for friendship, possibly more.

This is very similar to the first ad. Refer back to the original coding scheme and fill in the codes for this case.

	Code
GENDER1	_____
AGE-AD1	_____
AGE-DES1	_____
LTR1	_____
SEXUAL IM1	_____
APPEAR-AD1	_____
APPEAR-DE1	_____
SECUR1	_____
ACTIVE1	_____

Let's see if your coding is accurate. You should have coded GENDER1 as 1; for AGE-AD1 the coding is also 1; AGE-DES1 is 3. Again the phrase "friendship, possibly more" leads us to code 2 for LTR1 and 3 for SEXUAL IM1. The advertiser provides a minimal physical description of himself (6', H/W proportionate) but does not stress attractiveness—APPEAR-AD1 should be coded as 2. He neither provides a physical description of the person sought (APPEAR-DE1 should be coded as 1) nor requests financial security (SECUR1 is coded 2). He does indicate an active lifestyle—code 1 for ACTIVE1. You now have completed coding for the first two cases. Return to the main menu and answer Yes to the prompt about saving changes to the data file.

When coders read and code a large number of cases, they unintentionally may begin to modify the meaning of the codes. This is known as coder drift—criteria used for the later cases drift from those used for the earlier cases. The more subjective the coding, the more likely this problem is to occur. In this example, coder drift is much more of a problem when coding sexual implications of the ad than when coding the gender of the advertiser. This is another reason for having detailed coding instructions that the coder often refers back to and for using multiple coders.

For the present, let's assume that the codings are reliable and do a preliminary test of our hypothesis that the content of the ads will differ by gender. But before we select an analysis task, let's have the category "not stated" temporarily changed to missing data. We've used this feature several times before, so it should be familiar to you by now. Select the [File Settings] option from the main menu. In one of the blank boxes, type **Not Stated**, and then click [OK]. In all subsequent analysis, cases coded NOT STATED on a variable will be assigned the missing data value on that variable.

Now let's see whether there is a difference between males and females in terms of whether they stated age requirements in their advertisements. We'll use the CROSS-TABULATION task to analyze this.

Data File:	**CONTENT**
➤ *Task:*	**Cross-tabulation**
➤ *Row Variable:*	**3) AGE-DES**
➤ *Column Variable:*	**1) GENDER**
➤ *View:*	**Tables**
➤ *Display:*	**Column %**

AGE-DES   by   GENDER
Cramer's V: 0.442 **

AGE-DES		MALE	FEMALE	TOTAL
	UNDER 40	30	10	40
		60.0%	20.0%	40.0%
	40 & OVER	3	15	18
		6.0%	30.0%	18.0%
	NOT STATED	17	25	42
		34.0%	50.0%	42.0%
	TOTAL	50	50	100
		100.0%	100.0%	

We can see that males are more likely to seek someone under 40 than are females. Remember, however, that we must consider the possibility of spurious relationships. Perhaps there is some other factor creating this relationship. An obvious possibility might be the age of the advertiser—perhaps the males are younger than the females, and when we control for the advertiser's age, there may be no difference. Let's try this analysis.

Data File: **CONTENT**

Task: **Cross-tabulation**

Row Variable: **3) AGE-DES**

Column Variable: **1) GENDER**

➤ Control Variable: **2) AGE-AD**

➤ View: **Tables (Under 40)**

➤ Display: **Column %**

AGE-DES   by   GENDER
Controls: AGE-AD: UNDER 40
Cramer's V: 0.305
Warning: Potential significance problem. Check row and column totals.

		GENDER		
		MALE	FEMALE	TOTAL
AGE-DES	UNDER 40	20	7	27
		64.5%	36.8%	54.0%
	40 & OVER	0	1	1
		0.0%	5.3%	2.0%
	NOT STATED	11	11	22
		35.5%	57.9%	44.0%
	TOTAL	31	19	50
		100.0%	100.0%	

Just a reminder, the option for selecting a control variable is located on the same screen you use to select the other variables. For this example, select 2) AGE-AD as a control variable and then click [OK] to continue. Separate tables for each of the AGE-AD categories will be shown for the AGE-DES and GENDER cross-tabulation. The first table includes only those advertisers who are under 40 years of age.

This first table includes only those cases who are under 40 years of age. Examine the differences between gender. Let's look at the next control table: those advertisers who are over the age of 40.

Data File: **CONTENT**

Task: **Cross-tabulation**

Row Variable: **3) AGE-DES**

Column Variable: **1) GENDER**

Control Variable: **2) AGE-AD**

➤ View: **Tables (40 & Over)**

➤ Display: **Column %**

AGE-DES   by   GENDER
Controls: AGE-AD: 40 & OVER
Cramer's V: 0.544 **
Warning: Potential significance problem. Check row and column totals.

		GENDER		
		MALE	FEMALE	TOTAL
AGE-DES	UNDER 40	9	2	11
		56.3%	9.1%	28.9%
	40 & OVER	1	8	9
		6.3%	36.4%	23.7%
	NOT STATED	6	12	18
		37.5%	54.5%	47.4%
	TOTAL	16	22	38
		100.0%	100.0%	

Click the appropriate button at the bottom of the task bar to look at the second (or "next") partial table for AGE-AD.

Unfortunately, there are not enough cases to adequately test this alternative. The largest group, males under 40, has only 20 cases. The other groups have 10 or fewer. With only 100 cases, we are limited to examining bivariate relationships. If we were seriously interested in this issue, we would need to obtain a much larger sample of ads.

Your turn.

LIST OF ADVERTISERS

Use these cases and the code sheet that appears at the end of this list when doing Exercise 11.

Codes used by advertisers: LTR=Long-Term Relationship C=Christian J=Jew W=White B=Black H=Hispanic A=Asian N/S=Non-smoker N/D=Non-drinker S=Single D=Divorced M=Male F=Female H/W=Height/weight D/D-free=Drug- and disease-free P=Professional

1. CHRISTIAN AND KIND
SWM, 23, Christian values, 6'1", physically fit, attractive, enjoys summer activities, beaches, conversation. Would like to meet a Christian female for friendship, possibly more.

2. CHRISTIAN IN SEATTLE
Attractive, Christian SWM, 28, 6', H/W proportionate, active person, financially secure, stable and healthy. Would like to meet Christian female, for friendship, possibly more.

3. LOOKING FOR MS. RIGHT
SWM 23, 6'3", 165 lbs., brown/blue, enjoys camping, hiking, outdoors, movies, quiet nights. Seeking SF, 19–25, with similar interests. N/S, N/D.

4. A CUT ABOVE
Handsome, caring SWM, 47, 5'0", 190 lbs., honest, intelligent, business owner, enjoys arts, outdoors, travel. Seeking fit, classy lady, comfy in jeans/high heels, for LTR.

5. HARD-WORKING MAN
Handsome DWM, 40, 6'1", blond/blue, beard, country gentleman enjoys outdoors, camp fires, barbecues. Seeks attractive country girl, 35–45, N/S, light drinker for LTR.

6. BLACK GOLD
SBM, 37, 6'2", handsome, fit, romantic, easygoing, fun-loving, liberal, enjoys movies, sports, the arts, travel, dancing. Seeking lady with similar interests. 5'4"+ for LTR.

7. DON'T ANSWER THIS
Unless you seek a WM, 35, blonde/blue, who likes romantic nights and the outdoors. Please be WP/HF.

8. ROMANTIC TURK
Handsome, loving and passionate DWM, 46, 5'6", 140 lbs., willing to commit to a LTR, with active, fit, N/S lady 33–45, into personal and spiritual growth.

9. CARING KENT MAN
Thoughtful, caring SM, 39, 6', 185 lbs., brown/blue, enjoys motorcycles, music, movies, etc. Seeking SF, 21–38, H/W proportionate with similar interests, for possible LTR.

10. CHEF IN THE HOUSE
Attractive, Christian SBM, 30, N/S, N/D, easygoing, nice person, seeks very romantic, caring and honest SF friend, 25–33, N/S, N/D.

11. I COULD BE YOUR MAN
Are you 29–44, professional H/WF and have Christian values?

Professional, athletic DWM, 42, hazel eyes, 185 lbs., seeks woman to pamper and adore.

12. HIGH ENERGY WOMAN
Work/play hard? Me too. SWPM, 38, 5'10", slim, handsome. Not a homebody. Love R&B music, dancing, biking, talking, laughing, cuddling. Seeking SWPF 30–40.

13. TRUE MAN FOR FRIENDSHIP
Sensitive, intelligent, honest, handsome SWM, 40, 5'10", 160 lbs. Seeking partner to share life's pleasures. Attractive, proportionate, adventurous WF, 35–45. Emotionally available, monogamous LTR.

14. RENAISSANCE MAN
Handsome, Eastside WD/PCM, 45, business owner, 5'10", handsome, brown/blue, N/S, N/Drugs, athletic, enjoys outdoors, travel. Seeking pretty WPF, 5'2"–5'7", H/W proportionate for romance/LTR.

15. IN KEY
Attractive and active SWM, 34, 5'11", healthy, outdoors type, loves to cook and garden, honest and considerate, open-minded with common sense, enjoys talking.

16. AM I FOR YOU?
SWM, 38, 165 lbs., N/S, N/D, likes
outdoors and indoors too. Seeking
SF, 20–30.

17. SEEKING SOMEONE SPECIAL
SWM, 32, 5'10", 170 lbs.,
brown/hazel, enjoys walks, camping, and most outdoor activities.
Seeking SWF, 28–35, with similar
interests. Kids OK.

18. AVID ADVENTURER
Cute, rugged, strong SWM, 31, 5'11",
175 lbs., college-educated, financially
independent, loves snowskiing, boating. Seeking slender woman of grace
and sensuality, with inquisitive mind,
who shares above interests. N/S,
D/D-free.

19. GENUINE NICE GUY
N/S, attractive, DWPM, 34, seeks
slender, attractive WPF, 25–40, who
is honest and sincere, no games, for
summer country concerts, weekend
getaways, and romance.

20. WARM AND CARING
Active, athletic, DWPM, enjoys hiking, X-country skiing, dancing, ten-
nis, and quiet times. Seeking similar
energy level, female, 40–50, N/S, to
share common interests now and
into the future.

21. TALL, ATTRACTIVE, CELIBATE
SWM, 41, N/S, N/D, N/Drugs,
seeks WF, N/S, 5'9"+, H/W proportionate for friendship and variety of
activities; biking, skating, walks,
talks, animals, dance, quiet times,
possible LTR.

22. IN SEARCH OF
a best friend first. You: WF, H/W
proportionate, attractive, pretty, sexy,
25–45. Kids ok. Me: a gentleman,
WM, in 50s, ready for LTR.

23. FRIENDS FIRST
Handsome, fun SWM, 50, likes
movies, dining out, outdoors. Seeks
attractive SWF, 44–50, N/S, light
drinker ok.

24. CHRISTIAN FRIEND
Let's be friends first. Native to
Seattle. Likes scenic hiking, country
fairs, swimming and adventure. I'm
44, never married, seeking attractive
SWF, late 20s to early 40s. Let's meet
over coffee.

25. BUSTING OUT OF MY SHELL
Sincere, honest good-looking SWM,
31, long dark hair, 5'11", 150 lbs., fit,
likes movies, cats, music, creative
ideas. Seeking attractive, slender
female, 23–26.

26. SEEKING SOMEONE SPECIAL
SWM, 32, 6'1", 185 lbs., seeks SWF,
25–35, who wants to have fun and
take a chance on enjoying life.

27. PERRY MASON UNDERSTUDY
Finds legal matters intriguing, kind
SWM, seeks compatible female, 45+,
with humor. Mutual interests may
include: piano music, art, horses, old
houses.

28. LOOKING FOR LTR
Old-fashioned, romantic SWM, 23,
5'8", 140 lbs., brown/brown, enjoys
bowling, golf, family and the beach.
Seeking attractive, outgoing, SW/HF,
21–26, for friendship, possible LTR.

29. FUN-LOVING CHRISTIAN
Handsome, positive DWM, 35, 6'2",
physically fit, creative and spontaneous. Seeking an attractive, fun-loving female, 25–36. Kids ok.

30. LONELY IN KENT
SWM, 26, seeks SF, 21–30, interested
in outdoor activities, long walks and
romantic evenings.

31. SEEKING OLDER WOMAN
Intelligent, good-looking SWM, 31,
long hair, 5'10", in great shape, honest, creative. Seeking maturity of
attractive, slim woman, 35–55,
130 lbs., who enjoys Christian and
secular music, specifically heavy
metal. Light drinker/smoker ok. Call
me!

32. SUMMER ROMANCE
SWM, 41, 6', 175 lbs., brown/blue,
wants summertime love with interesting lady. I have good sense of
humor, very open, honest. Are you
adventurous?

33. NEW TO WASHINGTON
SWM, 30, funny, open-minded and
off-beat, not into looks, age or
money. Seeking female companion
for fun and romance.

34. LONELY AND SEARCHING
SHCM, 22, 5'5", 120 lbs., seeks
SW/HCF, 18–21, with long brown
hair, around 5'5" 139 lbs., who enjoys
Christian and secular music, specifically heavy metal. Light
drinker/smoker ok. Call me!

35. VERY HAIRY GUY
SM, 5'11", 215 lbs., long curly black hair, moustache and hairy body. N/D, N/S, likes mountains, movies, walks, rain. Seeking friendship with female, 35–47.

36. NEED TOUR GUIDE!
Attractive SWM, 32, financially secure. If you like intelligent conversation, dining out, and seek honesty and respect, I won't disappoint you! Seeking attractive, fun, responsible S/DF, 24–35.

37. SINGLE DAD
DWM, 43, 5'9", 185 lbs., loving father of two, handsome, enjoys camping, softball, boating, travel. You: S/DWF, 30-40, N/S, H/W proportionate, caring, honest, spontaneous.

38. TIME'S A WASTIN'
DWPM, 30s, 5'10", H/W proportionate, loves the outdoors. Seeking marriage-minded lady, under 35, N/S, who likes children and appreciates family life and values.

39. RENTON MAN
SWM seeks D/DWF, 29–37, H/W proportionate, who enjoys movies, talking, going out, dinners, etc. N/Drugs, N/D. Smokers ok. Friendship first.

40. LAKESIDE LIVING
DWM, 37, 6'3", fit and trim, N/S, romantic gentleman, jeans or suit, needs compatible south-end type for life's adventures. Kids ok.

41. ROMANTIC COUNTRY BOY
Me: 22, brown/brown, H/W proportionate. You: 22–30, blonde or brunette, H/W proportionate, likes C/W music. Seeking friend and romance for fun. No games.

42. BARBLESS IN SEATTLE
SWM, 39, 188 lbs., H/W proportionate, handsome, N/S, light drinker, loves fly fishing. Seeking SF, attractive, 22–39, H/W proportionate. Must also love fly fishing! Wants LTR!

43. CALL ME NOW
SWCM, 29, 5'10", 190 lbs., brown/brown, enjoys C&W music, movies, camping, sports. Looking for that one special woman to spend time with. N/S.

44. FRENCH MAN SEEKS AMOUR
SWM, 37, N/S, N/D, H/W proportionate, dark brown/green, seeks woman, 27–40, to travel and to explore with. Must be able to enjoy life.

45. SEEKING FUN AND YOU
DWM, 42, 6'3", medium build, enjoys fun, holding hands, walks, travel, camping, boating. Seeking light smoker/drinker. Pinochle a plus. LTR.

46. TOMBOYS AND SINGLE MOMS
Do you like the outdoors, art, tequila, rock music, boating, Mexican food, dancing, romance, adventure, laughter? SWM, 36, N/S, Mercer Island.

47. CLASSY, GQ GENTLEMAN
I'm 46, 5'10", trim, degreed, score an 8 out of 10, enjoy the finer things in life, honest, loving, gentle, and want to love a sweet, beautiful woman, 35–42.

48. ENDANGERED SPECIES . . .
committed father. WM, 43, 5'11", 220 lbs., seeks H/W proportionate mother, N/S, N/D, loves life and outdoors, most of all—complete and committed family life together. South King County.

49. TRUE BLUE CAMPER
SWM, 36, 5'8", 160 lbs., blonde/blue, enjoys hunting, camping, boating, light drinker/smoker. Seeking SWF, 28–40, H/W proportionate, for fun, romance, possible LTR.

50. EXTREMELY NICE GUY
Professional DWM, 47, 5'11", H/W proportionate, kind, considerate, likes drives, bike rides. I am a good catch. Seeking petite, professional female, 35–45, LTR.

51. COURT ME WITH ROMANCE
and experience the unforgettable, SWPF, 30, intoxicating elixir of sensuality and beauty, athletic, adventurous, seeks finely tuned, extraordinary SPM who can accommodate my passion for life.

52. COUNTRY GIRL/CITY STYLE
Cute, petite, DWPF, 37, non-Barbie, seeks tall, non-Ken, must love kids, animals, and fun. N/S, light drinker. Enjoys music, travel, warm nights, no couch potatoes.

53. AGE 22–27?
SCF, 21, pretty, enjoys outdoor activities. Seeking SM with a personality full of character, fun, and depth! Must be Christian, attractive, N/S, N/D.

54. BLONDE, BLUE-EYED
Educated DWF, enjoys beautiful music, dancing, outdoors, seeks DM gentleman, 45–55, who is kind, active, good conversationalist and happy.

55. HIGHLY EDUCATED BEAUTY
Tall, exciting, athletic, natural brunette seeks 4-year degreed (or more), charming business exec, 35–45, 6'+, interested in sports, the outdoors, cultural pursuits and a LTR.

56. ARTISTIC/ACTIVE/ ATTRACTIVE
Petite lady, 59 years young, looks 15+ years younger due to holistic lifestyle, seeks sensitive, secure soul mate, N/S, trim, fit gentleman of intelligence, awareness and sparkle.

57. BIG BEAUTIFUL SWF
Redhead seeks rugged, outdoor type, smoker ok, must like cuddling and dancing. Prefer over 5'11".

58. REDHEAD SWF SEEKS
sincere relationship with tall (6') SWM, enjoys camping, dancing and movies, who wants to have fun. Prefer N/S. N/Drugs, light drinker ok.

59. LOOKING FOR WHAT?
Let's look together. DWF, 50, 5'1", H/W proportionate, N/S, light drinker, family-oriented, likes having fun doing almost anything. Let's have coffee, talk, see what follows, maybe just friends.

60. LIFE WOULD BE COMPLETE
if I had the special someone to share it with. SWF, 35, blonde/blue, H/W proportionate, easy on the eyes, adventurous, passionate, playful, searching for a man that is open-minded and into sharing and caring as I am.

61. QUEEN SIZE
SWF, 27, seeks king size SWM, with good heart. Are you tired of being overlooked because of your weight? Me too, let's talk. Don't be shy!

62. GOLF ANYONE?
Outgoing, classy, athletic. Eastside DWPF, 44, brunette, H/W proportionate, financially secure, loves golf, biking, skiing, theater, fine dining, good conversation. Seeking professional counterpart, 38–48.

63. PRECIOUS JEM
JF, 40, 5'7", 180 lbs., brown/brown, sensual, sensitive, stubborn, enjoys crafts, country living, hobbies, traveling, seek SCM gentle-spirited, 35–43 with similar characteristics for friendship/LTR.

64. PASSIONATE AND STYLISH
Sensual intelligence, sublime touch, and the rhythm of the blues. All with significant sophistication at 31, for an appreciative professional over 45.

65. OLD-FASHIONED LOVE
DWF, 45, N/S, seeks LTR/possible marriage with professional M, with intelligence, integrity, humor, class. Only those commitment-minded please apply.

66. SMART, PERKY WITH ZEST
SWF, H/W proportionate, attractive, east and west coast person, with zest for life, likes boating, outdoor activities, every day is a joy. Seeking professional gentleman, 50s, with same ideas/values.

67. SEEKING A SPECIAL FRIEND
SWF, 47, H/W proportionate, enjoys fishing, camping, outdoors, animals, C/W. Seeking one-woman man, 43–55, light smoker/drinker ok, for friendship first, possible LTR later.

68. A FANTASY?
SWF, 43, 5'6", H/W proportionate, financially secure, active, enjoys life. Seeking SWM, 40–55, H/W proportionate, fit, financially secure, who can dance, N/S, light drinker, disease-free.

69. LEGS, LEGS, LEGS
SWPF, 30, 6'2", slender, looking for you . . . SWM, 28–40, N/S, who likes country music, dinners out, movies, picnics, and finding the humor in life.

70. SEEKING ROMANTIC COMPANION
DWF, 45, pretty, H/W proportionate, 5'4", N/S, light drinker, enjoys soccer, movies, cooking, cuddling, seeks SWM, 40–50, 5'10"+, attractive, fun-loving, H/W proportionate, sincere.

71. SPIRITUALLY-MINDED
SJPF, 42, pretty, with fun sense of humor, varied interests, H/W proportionate, N/S, light drinker, would like to get to know genuine male 40–48, under 6', for LTR. No games.

72. I'M A BRAT!
SWF, 29, wants to play. Love to
camp, golf, bike, travel. Seeking
SWM, 27–45, who is financially
secure, N/S, can keep with my quick
wit. No geeks.

73. LIFE'S BEST WHEN SHARED
Beautiful brunette, New Yorker CEO,
35, 5'4", 105 lbs., caring, intelligent,
Kosher, seeks handsome, successful,
educated, traditional JM, with strong
values; kind, generous, loving.

74. I'M LOST—FIND ME!
DWF, 30, loves sun, dancing,
romance, flexible—I'll follow your
lead. If you're fit and built like Van
Damme, you're my man. No geeks.

75. EASTSIDE CLASSY
Passionate, 39-year-old SWPF, enjoys
outdoors, very fit and versatile.
Seeking N/S, light drinking, tall, sincere and financially secure WPM for
LTR. I'm 5'10", blondish/blue, slender, romantic, pretty, fun!

76. ROCKY MOUNTAIN HIGH
High-energy, attractive SWF, H/W
proportionate, professional with passion for conversation, cuddling, sail-
ing, skiing, boating, bicycling, travel
and tenderness. Seeking SWM, 55+,
H/W proportionate, N/S.

77. R U MAN ENOUGH?
DWF, 38, 5'11", H/W proportionate,
enjoys biking, gardening, travel and
darts. Seeking tall S/DWM
employed, clean-cut, for LTR. Light
smoker/drinker, D/D-free.

78. BORN COUNTRY
A heart full of love, a genuine friend,
a happy lady! DWCF, N/S, N/D.
Very active, secure, slender, young 53.

79. SUMMER FUN
Attractive WF, 39, 5'10", blonde,
devoted part-time mom, seeks tall,
Eastside gentleman. We're athletic,
educated, healthy, N/S, desire LTR,
with romance.

80. RSVP ASAP
SWF, 25, easygoing, secure, professional, enjoys boating, scuba diving
and traveling. Seeks SWM, 25–35,
stable, attractive and financially
secure.

81. KIND & COMPASSIONATE
Petite, attractive, SWPF, early 50s,
seeks soulmate for LTR. I'm active,
spiritual and enjoy outdoor activities, movies, bookstores, travel, animals, and many other interests.

82. TAKE A CHANCE
Very attractive SWF, mid-30s, 5'6",
H/W proportionate, blonde/blue,
educated, outgoing, introspective,
caring, sensual. Seeking same in fun,
attractive male, 35–40, 5'10"+, 200 lbs.
for possible LTR. N/S, light drinker.

83. CUTE, SWEET & PETITE
Attractive DWF, 43, 5'2", 115 lbs.,
long light-brown/green, enjoys boating, hiking, walks on beach, trying
new things. Seeking someone special
36–47, 5'9"–6'2", physically fit, secure
for monogamous LTR.

84. COUNTRY RAISED
SWCF, H/W proportionate, 5'6", 160

lbs., country, lives in city. Seeking
SWCM, 40–48.

85. TOTALLY UNIQUE
This SWF, 42, attractive, bright, educated, unpretentious, enjoys country
living, nature, walks, animals, gardening, reading, relaxing. Seeking
SWM, similar age and interests, for
LTR.

86. BLUE JEANS & MINK
DWCF, 5'7", 130 lbs., fit, attractive,
family-oriented, spiritually/emotion-
ally strong with varied interests:
Mozart to camping, etc. Seeking honest WPM, 40–48 for LTR.

87. ATTRACTIVE BROWN-EYED
GIRL
This 40-year-old, H/W proportionate
N/S is looking for my best friend,
who wants LTR. Must be confident,
loyal, honest, emotionally/financial-
ly together, with Christian values.
Please call. N/S only.

88. CRITIQUE THIS
SWF, 27, outgoing, spontaneous, professional, enjoys boating, jeeping,
island hopping. Seeks SWM, 27–37,
attractive, confident, financially
secure, adventurous, with similar
likes.

89. SMILING EYES
Attractive DWF, 47, N/S, light
drinker, N/Drugs, outgoing, affectionate, kind, spontaneous, positive,
active, professional seeks SWM,
43–50, H/W proportionate 6', with
similar values and morals. Loves
nature.

90. ROSES ARE RED
Violets are blue. I'm sick of the dating scene, how about you? SWF, 27, sick of game players, losers. Seeking LTR with professional SWM 27–35.

91. FASHIONABLY LATE
SWF, 39, 5'7", athletic, attractive, seeks one special man, intelligent, 6'+, successful, adventurous. Looking for somebody special.

92. WARM, COZY LIFE
DWF, 59, has happy, affectionate, tasteful lifestyle to share with sturdy, appreciative, educated man. Honesty, humor, communication, compassion a must for both. Eastside, Seattle, NS.

93. ENERGETIC
Sophisticated, educated lady, likes outdoors, ethnic activities, sunshine, flowers, life in general. Life's too full for movies or TV. Seeking SWM, 45–55, 5'10"+.

94. LIFE IS GOOD . . .
and getting better! Widowed WF, 53, enjoys family/friends, home/garden, travel/walks. Seeking articulate, metaphysical, romantic SWPM, N/S, light drinker. N/Drugs. Leave your message.

95. ALONE
Attractive DWF, 50+, wants companion/friendship with spontaneous, adventurous S/DWM, 55–60, 5'10"+, N/S, light drinker, with good sense of humor, enjoys outdoors and travel.

96. CUTE, FUN AND JUST ME
Honest, fun-loving, brown/brown, 33, 5'4", H/W proportionate, D/D-free, N/S, light drinker, seeks same in Harrison Ford/Tom Hanks type to share life's pleasures.

97. PETITE BLONDE
Successful, professional DWF, 44, Christian, N/S, light drinker, "loves water & sun," pretty smile, 5'3", 110 lbs., seeks successful man who loves children and family for fun, romance, LTR.

98. WILL PAINT YOUR WAGON
Scintillating artist, 60s beauty. WPF, stylish fun, gourmet cook, desire savvy SWP "He-man," 55–65, 5'9", N/S, fun loving companionship/life adventures.

99. CARING AND HAPPY
DWF, 54, 5'6", N/S, height/weight proportionate, enjoys golf, movies, dancing, sports. Happy with life, like to meet man with similar interests. Let's meet for coffee.

100. SEEKING PRINCE CHARMING
42-year-old mother (of three boys), 5'2", long blonde/brilliant blue, well-educated, active, altruistic. You? Fun, adventurous, secure, no hang-ups. Integrity, affection, kindness and prayers are musts.

CODE SHEET FOR CODING NEW VARIABLES

Case	SEXUAL IM1	SECUR1	Case	SEXUAL IM1	SECUR1
1			26		
2			27		
3			28		
4			29		
5			30		
6			31		
7			32		
8			33		
9			34		
10			35		
11			36		
12			37		
13			38		
14			39		
15			40		
16			41		
17			42		
18			43		
19			44		
20			45		
21			46		
22			47		
23			48		
24			49		
25			50		

NAME:

COURSE:

DATE:

EXERCISE

11

Workbook exercises and software are copyrighted. Copying is prohibited by law.

WORKSHEET

This worksheet section continues to use the CONTENT data file. If you did not immediately continue from the previous section, you will need to open the file and set the "not stated" category to missing data. To do this, select the FILE SETTINGS option from the **FILE & DATA MENU**. In one of the blank boxes, type the phrase **Not Stated** and then click [OK].

1. Ideally, all variables on all cases would be coded at least twice. However, this type of coding takes considerable time. To understand the process, you will code (enter data for) the *first 50* cases on just two variables: 14) SEXUAL IM1 and 17) SECUR1. Look at the special notes on each of these variables in the variable descriptions before beginning your coding. Use the code sheet preceding this worksheet section to code the cases first, then enter the data into the MicroCase data file.

    Select the ENTER DATA task from the **FILE & DATA MENU**. Click [OK] to accept the default settings as Grid Entry format.

    a. After you have finished the coding, use the CROSS-TABULATION task to look at the relationship between 5) SEXUAL IMP and 14) SEXUAL IM1.

    How many cases have been coded the same on both
    variables?   _____

    $V =$ _____

    Based on this information, do you think 5) SEXUAL IMP is reliable
    enough to be used in analysis? (Circle one.)   Yes   No

    b. Now look at the relationship between 8) SECUR and 17) SECUR1.

    How many cases have been coded the same on both
    variables?   _____

    $V =$ _____

    Based on this information, do you think 8) SECUR is reliable
    enough to be used in analysis? (Circle one.)   Yes   No

2. When a range is given rather than an age, the younger end is used in the coding. What other technique(s) could be used for coding the age when a range is given? Discuss the relative advantages and disadvantages of these approaches.

3. Test the hypothesis that the sexual implications are more likely to be stressed in ads placed by males than in those placed by females. Please show your analysis.

Data File:	**CONTENT**
Task:	**Cross-tabulation**
➤ *Row Variable:*	**5) SEXUAL IMP**
➤ *Column Variable:*	**1) GENDER**
➤ *View:*	**Tables**
➤ *Display:*	**Column %**

	Male	Female
Explicit	_____	_____
Implicit	_____	_____
None	_____	_____

V = _____

Prob. = _____

What is your conclusion?

4. Test the hypothesis, using the original variables, that financial security of the desired respondent will be more important in ads placed by females than in those placed by males.

	Male	Female
Stated	_____	_____
None Ind.	_____	_____

V = _____

Prob. = _____

What is your conclusion?

5. Test the hypothesis that an interest in long-term relationships is more likely to be expressed in ads by females than in ads by males.

	Male	Female
LTR	_____	_____
?	_____	_____
No	_____	_____

V = _____

Prob. = _____

What is your conclusion?

6. Test the hypothesis that an interest in long-term relationships is affected by the age of the advertiser.

	Under 40	40 & Over
LTR	_____	_____
?	_____	_____
No	_____	_____

V = _____

Prob. = _____

What is your conclusion?

7. Test the hypothesis that an active lifestyle is more likely to be stressed by men than by women.

	Male	Female
Yes	_____	_____
No	_____	_____

V = _____

Prob. = _____

What is your conclusion?

Projects

CREATING A MICROCASE FILE

With Student MicroCase, you can create a file with 50 variables for as many as 100 cases. While you can create only one such file at a time, you can create new files as many times as you want—each new file simply replaces the old file. If you do any of the projects described later, you will need to set up such a file.

Let's use a very small set of hypothetical data to demonstrate how to create a MicroCase file and enter data. We will use just three variables for just five cases, but the same procedures would be used if you were creating a much larger data file.

Suppose we have five students who answered the following mini-questionnaire:

1. Do you live at home, in a dorm, or where?

   1. At home with my parent(s)
   2. In my own apartment or house
   3. In a dorm
   4. In a sorority
   5. In a fraternity

2. I'm not sure that college is worth all the bother.

   1. Strongly agree
   2. Agree
   3. Disagree
   4. Strongly disagree

3. What is your GPA (Grade Point Average)? _____ (write in number)

Each answer to the first two questions has been coded with a number. This is not necessary in the last question, since the response itself will be a number. Generally, when you are entering data for a survey, you will have a questionnaire for each respondent and you would enter data directly from the questionnaire. For example, the responses of the first student are shown below in bold.

1. Do you live at home, in a dorm, or where?

   1. At home with my parent(s)
   2. In my own apartment or house
   3. In a dorm
   4. In a sorority
   5. In a fraternity

2. I'm not sure that college is worth all the bother.

   1. Strongly agree
   2. Agree
   3. Disagree
   4. Strongly disagree

3. What is your GPA (Grade Point Average)? ____**3.6**____ (write in number)

Hypothetical data are provided in the table below for the first five students who were surveyed.

	LIVE?	COL.WORTH	GPA
Student 1	1	3	3.6
Student 2	3	2	3.2
Student 3	2	4	4.0
Student 4		1	2.3
Student 5	1	4	3.0

We go through three steps in order to create a new MicroCase data file and enter the student responses: creating the file, defining the variables that will be included in the file, and entering the actual data for those variables. Let's go through these steps for the student data.

STEPS IN CREATING A MICROCASE FILE

STEP 1: CREATE A NEW DATA FILE

Important Note: Before you learn how to create a data file, it is important to know a couple of things. First, this version of Student MicroCase permits you to create only one data file at a time. To create a second data file, you must, essentially, replace the first data file you created. Next, if you want to replace a data file you created, DO NOT use the delete key or the recycling bin (or any other Windows operation) to delete this file. This file must be *replaced* within Student MicroCase using the [New File] option. For more details on this, see the last part of the instructions below entitled "Replacing a File That You Created with a Newer File."

Go to the **FILE & DATA MENU** in Student MicroCase. Select the task named NEW FILE on the menu. On the screen that appears, you will be asked to provide a name and a description for the file that you are creating. For the File Name, type a descriptive name consisting of 1–8 characters. For this example, type **STUDENT** and then press the <Tab> key to move the cursor to the File Description box (if you accidentally pressed <Enter>, you will be asked if you want to enter a variable description). Then type a description of the file, such as the following: **Student Questionnaire**. Then click [OK] to return to the main menu.

STEP 2: DEFINE THE VARIABLES

Each question will become a variable in this data file. So now you will provide information about each variable (or question). Select the DEFINE & EDIT VARIABLES task. The term *undefined* is highlighted. The program is asking for a name for the first variable. *A variable name contains from 1 to 10 characters, including any blank spaces in it.* The name should give you a quick idea of what the variable is. The first question asks where the student lives. Let's call this variable "LIVE?". So where the term *undefined* now appears, type **LIVE?** and press the <Tab> key.

Next, give the variable a description. For this example, type in the first question: **Do you live at home, in a dorm, or where?** and then press the <Tab> key.

The program now asks for the minimum possible value and the maximum possible value for the variable. For the minimum value, type **1** and press the <Tab> key. For the maximum possible value, type **5** and press the <Tab> key. Press the <Tab> key again to skip the question about the number of digits.

Now you can give a 1–10 character label for each value (or category) of the variable. In this situation, code 1 simply stands for "at home with my parent(s)." So type **At home** in the Category Name column, and then press the <Tab> key. For category 2, type **Own place** and press the <Tab> key. For category 3, type **Dorm** and press the <Tab> key. For category 4, type **Sorority** and press the <Tab> key. Finally, for category 5, type **Frat** and press the <Tab> key.

At this point, check the information on the screen for accuracy. If there are any mistakes, you can simply position the mouse cursor wherever you need to in order to correct the mistake.

Next, click the [Continue] button at the top of the screen in order to start work on the second variable. The second variable is a question about the value of college. Let's call this variable "COL.WORTH". So, type **COL.WORTH** and press the <Tab> key. For the variable description, type **I am not sure that college is worth all the bother** and press the <Tab> key.

Next, enter **1** for the minimum value and **4** for the maximum value. Press the <Tab> key again to skip the question about the number of digits. Then type the category name **Str Agree** and press the <Tab> key. Next, type the category name **Agree** and press the <Tab> key, then **Disagree** and press the <Tab> key. Finally, type **Str Disagr** and press the <Tab> key.

Click [Continue] to begin work on the third variable, the Grade Point Average. Type **GPA** and press the <Tab> key. For the variable description, type **Grade Point Average** and press the <Tab> key.

Unlike the previous variables, this is not a categorical variable and there is a decimal point in it. Further, in dealing with such variables, it might be difficult to determine in advance what the minimum and maximum values are. So, press the **[Tab]** key twice to skip the minimum and and maximum values. For the **Decimal Digits**, type **1** to indicate that there is one place to the right of the decimal point.

We have now defined all the variables. So, click [Menu] to exit from this screen and return to the main menu.

STEP 3: ENTER THE DATA

You are now ready to type in the actual data. Select the ENTER DATA task, select the [Grid Entry] option, and click [OK].

The next screen presents the data entry matrix. Click on the cell at the upper left to position the cursor there. Refer back to the hypothetical data for the five students shown earlier in the table. The first student lives at home (category 1). So, type **1** and press the <Enter> key. He, or she, selected disagree (category 3) as the response to the second question. So, type **3** and press the <Enter> key. The student has a GPA of 3.6. So, type **3.6** and press <Enter>.

We're now ready for the second student. The answer to the first question is 3, so type **3** and press <Enter>. Type **2** for the second variable and press the <Enter> key. Type **3.2** for GPA and press <Enter>.

Although the third student lives in his or her own apartment (category 2), type **6** and press <Enter> to demonstrate a point. The program gives a message that the value is out of range—the value is outside the range defined by the lowest and highest values that we specified. This feature helps to avoid at least certain types of data entry errors. Click [OK] to clear the message, click on the first cell for case 3, and enter **2**. Then enter **4** as the response to the second question and **4.0** as the GPA.

We are now ready for the fourth student. Notice this student did not answer the first question. Simply press <Enter> to leave this variable blank for this case and the program will automatically treat this value as missing. For the COL.WORTH variable, enter **1**, and then enter **3.0** for GPA.

Do not enter any data for the fifth and final case—we will do that in the next section. For now, click on [Menu] to return to the main menu. The program will now ask whether you want to save the changes to the data. Click [Yes]. That's it. You have created a complete data file, and you can now use it the same way that you use any of the other data files in the MicroCase package.

Question Entry

This option will show the question with its possible answers during data entry. The value for the current case is selected by clicking on the appropriate category.

Those inexperienced in data entry are less likely to make mistakes using this option.

Using the STUDENT file you created in the previous section, select the ENTER DATA task from the **FILE & DATA MENU**. Then select the [Question Entry] option, and click [OK]. When the data entry screen appears, the cursor is shown in the grid on the next case and variable. Let's enter the fifth case to our hypothetical data file. First, it may be necessary for you to use the scroll bars located to the right of the grid to position the highlight at the first variable for the fifth case. Go ahead and do that if you need to.

The fifth student lives at home so double-click the **At home** category shown in the entry box at the top of the screen. The value 1 will automatically appear in the appropriate location on the grid below, and the entry screen will advance to the next survey question. This student strongly disagrees with the statement that "I'm not sure that college is worth all the bother" so double-click on the **Str Disagr** category shown at the top of the screen. The GPA variable doesn't have predetermined categories from which to select (because the values will be decimal), so you are prompted to enter a value. Enter **3.0** and press <Enter>. At this point you are advanced to the next case in the data file. We have finished data entry of our five hypothetical cases, so you can return to the main menu by clicking the [Menu] button. You will be asked if you want to save your data—click [Yes].

Rekey Verification

In order to eliminate data entry errors, you might want to reenter the data using the rekey verification process. The underlying assumption of this validation procedure is that the same entry error is unlikely to occur both times a value is entered.

With the STUDENT file open, select the **FILE & DATA MENU**, select the ENTER DATA task, select the [Enable Rekey Verification] option and one of the format options for data entry, and click [OK]. Now you can reenter the data for each case and each variable. If the value matches the existing value, the program continues in the normal fashion. If there is a mismatch, the program will ask you to select the correct entry.

Return to the main menu by clicking the [Menu] button. You will be asked if you want to save your data—click [Yes].

MODIFYING A FILE THAT YOU CREATED

Once you have created a MicroCase data file, you might need to modify it for various reasons (e.g., to correct errors, to add more data, or to add new variables).

Adding More Data or Changing the Data.   If you are correcting data or adding new data (or eliminating some of the data already in the file), then you begin by opening the file and selecting the ENTER DATA task. Select either data entry option and the matrix of data will then appear. You can move about in the data by simply clicking on the spot to where you want to go. You can make any changes you need to make. When finished, click on [Menu] to return to the main menu.

Modifying the MicroCase Data File. If you are modifying the MicroCase data file (e.g., adding new variables or correcting errors in the variable descriptions), then you begin by selecting the DEFINE & EDIT VARIABLES task. The screen that appears is ready to accept new variables, and you simply continue as before if this is what you are doing. If you are making changes on old variables, then use the arrow buttons that appear below the **Current Variable Number** to scroll to the variable you want to change. For example, if you were making changes to variable 2, then you would use the left/right arrows underneath the box to scroll to the number 2.

When you finish with any additions or modifications, you click the [Menu] button to return to the main menu.

REPLACING A FILE THAT YOU CREATED WITH A NEWER FILE

If you have gone through the preceding example, then you have created a data file named **STUDENT**. Having completed this exercise, however, you are now ready to create a file of your own. How do you replace the STUDENT file with a new file? You simply create a new file (starting with the NEW FILE option) and it replaces the old file automatically. *Do not use the <Delete> key to remove the existing file.*

The last point deserves extra emphasis. As noted at the beginning of the "Create a New Data File" section, this version of Student MicroCase permits you to create only one data file at a time. To create a second data file, you must replace the first data file you created using Student MicroCase's NEW FILE option. DO NOT use the delete key or the Windows recycling bin (or any other Windows option for deleting files) to remove this file. If you do, the NEW FILE option will no longer work in Student MicroCase.

If, despite the above warnings, you accidentally delete a file you created, there is a way to reset this file so that the NEW FILE option works again. First, close the Student MicroCase program. Then use the My Computer feature in Windows to locate the data files (e.g., GSS, USA) for Student MicroCase. If Student MicroCase is accessing your data files from the floppy diskette (such as in a lab setting), then you will find these files on your A or B drive. If you did the standard installation and you installed the file to your hard drive, you will find these files in the C\Program Files\MicroCase\CS directory or folder. Once you have found your data files using My Computer, double-click on the application file named RESET. A brief program will run that resets part of Student MicroCase so that you can again create a data file. Answer **Y** (for Yes) to the prompt warning that a file will be overwritten. You are now returned to the Window desktop (you may have to click the "x" to first close the DOS box that appears when you run the RESET program). Then restart Student MicroCase.

HUMAN SUBJECTS GUIDELINES

You have probably read about the medical experiment in which men with syphilis were denied penicillin, a known cure, in the quest for additional scientific information. Or the studies of the effect of radiation on individuals who were not informed of the potential risks. While the actual research studies in these incidents undoubtedly are more ambiguous ethically than these headlines suggest, the fact remains that until relatively recently, the ethical treatment of subjects in research was left almost exclusively to the judgment of the researcher. In the 1970s, to protect the public from unethical research, granting agencies started to require that all research must stay within a set of ethical guidelines. To guarantee conformity, each research proposal must be approved by a special committee. Today, virtually all universities and granting organizations require a *human subjects* review for proposed research in which the subjects or participants are humans. (Incidentally, these procedures also protect researchers from lawsuits by unethical subjects.)

Educational projects, such as those suggested in this section, are generally exempt from this process unless the results of the research will be published. Your instructor will inform you if you will need any special clearance before proceeding with an assigned project.

If, at some future time, you wish to conduct your own research, you should review the ethical concerns described in the text for your particular research approach. In addition, you should design your research project using the following suggestions:

- If possible, obtain the informed, voluntary consent of subjects in research. If it is not feasible to obtain consent in advance of the research, then try to obtain consent afterward.

- If there is even a modest degree of risk of harm to the subjects, then it is mandatory that you obtain the informed, voluntary consent of subjects prior to the research.

- Minimize any risk of harm to research participants which might result from the research in any way—from the initial stages of the research to any problems that might occur in the aftermath of the publication of results.

- Take whatever steps are necessary to protect the identity of research participants (e.g., destroying information that might identify participants, disguising the location where the research took place, and presenting results of the study in such a way that no individuals can be identified).

- Do not deceive the subject unless it is the only feasible way to achieve the research objective and adequate provisions have been made to protect the subject from harm.

- Protect the privacy and dignity of the participants.

- When subjects have been used in experimental research, detect and remove any harmful consequences to the subject (e.g., stress).

You may also want to consider the following three aspects, which come under particularly close scrutiny during evaluation of the use of human subjects in social science.

1. INVASION OF PRIVACY

This is probably the most common ethical problem in social science research. Before collecting any data, social researchers must inform each participant who will have access to the information and how the information will be used. Steps must then be taken to maintain the promised degree of privacy. This is primarily an issue in survey research, and the text discusses several methods used to maintain the confidentiality of such information. In field research, you must also be careful to protect the privacy of those observed.

When illegal or criminal behavior is studied, the researcher has an additional burden. Social scientists, like newspaper reporters and unlike doctors and lawyers, have no right of confidentiality regarding information and sources. If a list of survey respondents or a tape recording is subpoenaed by a court, a researcher may be held in contempt and punished for refusing to produce the information. In such studies, the researcher should not promise a greater degree of confidentiality than is legally possible. A related problem occurs if the researcher uncovers evidence of continuing criminal activity. For example, a field researcher may find that a particular gang burns buildings on a regular basis and may even have knowledge of future crimes. Failure to report this information could lead to criminal charges against the researcher.

2. STUDIES INVOLVING DECEPTION

In some studies, subjects or participants may not be told the complete truth. This is most common in experimental studies. For example, the experiment described in the text involves a deception: Subjects are led to believe they are listening to other participants when in fact they are listening to a tape recording and, moreover, the taped conversation encourages them to believe that another participant has suffered a seizure. Other types of research may also involve some level of deception. Field researchers may sometimes imply they hold a position when, in fact, they do not. For example, in observing a hospital, a researcher may dress as an orderly, a nurse, or a doctor to be less obtrusive. Even survey researchers may sometimes wish to conceal the exact purpose of their survey.

Deception is a touchy ethical issue. If the same information can be obtained without deception, this is almost always preferable. Most human-subjects committees will weigh the degree and effect of the deception against the potential gain of the study. If deception is used, subjects must be *debriefed* after the study to minimize any consequences of the deception. For example, in the experiment described above, subjects who did not leave the room to seek help may later worry and feel guilty about their behavior. The researchers are responsible for discussing the actual study with subjects so that these feelings do not occur. Unfortunately, no amount of debriefing can erase the subject's knowledge of how he or she behaved in the particular circumstances.

3. INFORMED CONSENT AND COERCION

Participants and subjects in studies must be informed of the purpose of the study and must be allowed the opportunity to refuse to participate. For example, an instructor cannot *require* any student in a class to complete a questionnaire for a research project—such participation must be voluntary.

When adequate information is provided, most social science research projects are readily approved by the relevant committees.

PROJECT 1: DOING A SURVEY

In this project, you can learn something about students at your college or university. Before you begin, decide on the purpose of your survey. Perhaps you want to see if males and females at your school exhibit gender stereotypes. Or you might want to see if there are sex differences in deviance: Are males more likely to have been picked up by the police than are females? You could determine if students at your school are well informed on certain demographic facts, such as what the world population is. You could see to what extent students' political views match those of their parents. You could even replicate the methodological experiment described in Exercise 10.

After you have selected your topic, design your questionnaire. A variety of possible questions are listed in Appendix B, or you can design your own questions. You can have up to 50 variables in Student MicroCase.

After you have developed your questionnaire, construct your analysis plan. Which questions will be independent variables and which will be dependent variables? What analysis will you use to answer your research question? Are there possible sources of spuriousness to consider? Any potential intervening variables? It is extremely important to develop your analysis plan before finalizing the questionnaire. This is the only way you can be sure to collect data on all relevant variables. **Warning:** Even very experienced professional researchers have made ridiculous errors of omission in their surveys. For example, a multimillion-dollar longitudinal study of education, following people from ages 16 through 26, failed to ever ask whether respondents had completed college. So, after you are sure you have included everything you will need to know, check again.

You will need to decide whether you will use phone interviews, face-to-face interviews, or mailed questionnaires. If you plan to conduct phone interviews, compose your opening statement. For example, you might say the following:

Hello, this is _____. I'm a social science student at _____.
As a class project, I'm conducting interviews with a sample of students.
Could you spare five minutes to answer some questions?

If you are going to do face-to-face interviews, you will want to work out your approach, perhaps adapting the opening phone statement. If you are going to mail the surveys, you will need to write an appropriate cover letter.

You can now pretest the survey on two or three friends. If necessary, rework the questionnaire. Again review your analysis plan to make sure that you have all relevant information.

Now you can select your sample. Obtain a list of phone numbers or addresses for all students (or perhaps all undergraduates) at your college or university, and randomly select a sample of 50 students. You can then proceed to collect the data.

After the data have been collected, create your MicroCase data set. Enter the data from the questionnaires. Complete your planned analysis and write up the results.

PROJECT 2: DOING A COMPARATIVE STUDY

1. In this project, you'll use information from the *Census of Retail Trade* to test ideas about the effect of the age profiles on retail trade. Before you begin, make sure that you have access to the *Census of Retail Trade*. You may want to examine these reports to get an idea of the type of information available.

Develop two hypotheses about the effect of the age distribution on the consumption of products and services, using states as the units of analysis. For example, you might speculate that states with a high percentage of children will have a higher ratio of spending in grocery stores to spending in restaurants.

After you have developed your hypotheses, you will need to find data on the variables. The age distribution is available from many sources—the *Statistical Abstract of the United States* is probably the most convenient. If necessary, convert this number to a rate, that is, from the number of children under age 12 to the percent of the population under age 12. You can obtain the measures of the dependent variables from the *Census of Retail Trade*. Again convert these to rates.

Next, create a MicroCase data file. (Refer to "Creating a MicroCase File" at the beginning of this section.) Test your hypotheses and write up your results.

2. Create a MicroCase data file of the 25 largest nations. (Refer to "Creating a MicroCase File" at the beginning of this section.) You will first have to determine which nations should be included. You may obtain population figures for each nation from the *Statistical Abstract of the United States*. Develop a hypothesis that you can test with this data set. For example, you might speculate that the higher the per capita income, the lower the birth rate.

Find appropriate data for each country. Convert the numbers to rates if necessary. Enter the data into your MicroCase data file and test your hypothesis. Write up your results.

3. Select a probability sample of 100 counties or 100 cities in the United States. Collect data to test hypotheses about factors that lead to higher crime rates in counties (or cities). The *County and City Data Book* would be a good source of data.

Convert the data to rates if necessary. Set up a MicroCase data file and test your hypotheses. Write up your results.

PROJECT 3: DOING A FIELD STUDY

Two suggested projects are described here. Since you are observing in public places, you need not obtain permission from those individuals you will observe. However, you will probably want to appear as a "natural" element of the setting. If people know you are taking notes on their behavior, they may not only change the behavior, but also confront you or file a complaint about you. An easy guise for a college student is to pretend to be studying: some reading, some writing, and much thoughtful gazing into space.

When you organize your notes and write your report, develop a central theme. What did you find most interesting? Use your observations to support your interpretation of events.

1. Observe children on playground equipment on several different days. Before you start taking notes, you will want to think about the types of questions you might be able to answer. Which pieces of equipment are most popular? Are these preferences affected by the age or sex of the children? Do children play in groups or by themselves? Are there any age or sex differences in group play? Are there pieces of equipment that encourage cooperation or promote competition among the children? Does the play tend to be segregated by age and sex? Are adults present? To what extent do the adults control the children's behavior? Are some children more popular than others? Are there any "outcasts"? If so, can you determine why this happens?

Remember that adults lurking around playgrounds could appear suspicious. If you plan to observe at a school playground, obtain permission from the principal in advance. You might ask the principal to sign a letter that you have prepared so that you have proper "credentials." If you are observing at a public park, you might want to have a letter from your professor on your college stationery.

2. Observe a fast-food restaurant at different times on different days. As a good field researcher, you should obtain permission from the manager or owner before beginning your observations. If you plan to observe the staff, ask the manager not to reveal your role. As a courtesy, try not to occupy a table when customers need it. If necessary, limit your observations to times when the restaurant is less crowded.

You might choose to observe the staff or the customers or both. What is the age and sex composition of the staff? Who's in charge? Does informal staff interaction appear to be a function of personal characteristics or of their role in the restaurant? Do there appear to be any conflicts among the staff? If there are, how are these conflicts resolved? Do the staff talk with customers or just fill their orders? What is the attitude of the staff? How do they react when inundated with customers? What do they do when there are no customers?

How does the customer base change over time and over the days of the week? How do individuals who eat by themselves behave? Do they ignore others, perhaps reading newspapers or books, or do they try to engage others in conversation? Are there any sex, age, or occupational differences in this behavior? What sort of people eat together? Aside from family groups, are most groups segregated by age and sex? Which kind of customer eats the fastest? Which kind takes the longest? Do some individuals use this as a social event, or do all customers appear to be there strictly for the food? How do they treat the staff? Are there age, sex, or occupational differences in this treatment?

PROJECT 4: DOING AN EXPERIMENT

1. In this experiment, you can assess the effect of the physical attractiveness of children on adults' responses to them. The independent variable will be photographs of children who differ in attractiveness. A written statement about the child will be attached to the photograph—this description will be the same for all photos. The subjects will then be asked to answer questions about the child.

The first step is to develop the experimental manipulation. Obtain pictures of many children. These should all be the same sex and same apparent age. Now have three or four persons sort the pictures in terms of attractiveness. Record the ranks for each photo. Select two photos—one ranked attractive by all raters and one ranked unattractive by all raters. These two pictures will now be the experimental manipulation.

Develop the description that will accompany the pictures. The following description was adapted from a newspaper account of a child available for adoption:

> Three-year-old *David* is a solemn little *boy* who seems to carry the troubles of the world on *his* shoulders. But *his* mellow, sweet personality makes *him* a favorite of everyone who meets *him*. *He* has an amazing ability to put together Legos and *he's* good at games. *He* loves animals and they seem to love *him*. *He* plays better by *himself* than in groups, though *he* does fine with other children. *He'd* do best as part of a family where *his* quiet spirit won't be lost. [Words in italics have to be changed if pictures of female children are used.]

Now determine how you are going to measure the dependent variable. You might say that you are trying to assess the effectiveness of this type of adoption description and ask the subjects to rate the child on several scales.

Will David be adopted soon?
1____Very likely
2____
3____
4____
5____
6____
7____Very unlikely

How well will David adapt to a new family?
1____Very well
2____
3____
4____
5____
6____
7____Not at all well

How mature is David for his age?
1____Very immature
2____
3____
4____
5____
6____
7____Very mature

Will David have problems in school?
1____Very unlikely
2____
3____
4____
5____
6____
7____Very likely

Be sure to include a question that checks on the effectiveness of the manipulation.

Please rate this child in terms of physical attractiveness:
1____Very unattractive
2____
3____
4____
5____
6____
7____Very attractive

Since you will not be trying to generalize to a population, you may select your subjects in any way that is convenient. You should randomly assign each subject to an experimental condition. You might do this by flipping a coin—assign those with heads to condition 1 and those with tails to condition 2. You should have about 15 subjects in each experimental condition. Simply stop assigning subjects to a condition after you have reached the desired number.

Ask each subject to look at the picture and read the description. You could then interview the subject and fill in the questionnaire yourself. Or you could have the subject fill in the questionnaire.

Then create a MicroCase data set and enter the data. Be sure to create a variable for the experimental condition—simply assign each subject a 1 or 2. You could then determine if there were differences in the dependent variables across the experimental conditions. Be sure to check that the experimental manipulation was effective—subjects perceived one child as more attractive than the other.

ALTERNATIVE EXPERIMENT: Do this same study using homely and attractive dogs.

2. Use similar procedures to replicate the study described in Chapter 6 in the textbook. That study examined the effect of a candidate's gender on voters' choices. In each condition, subjects were asked to read a description of a candidate. They were then asked a series of questions about the candidate. The only difference between the two conditions was the name of the candidate. In one condition, the candidate was clearly female and, in the other condition, the candidate was male.

First, create a description of a candidate. Since you want subjects to be aware of the gender of the candidate, you might use personal pronouns (he, his, him, she, hers, her) liberally. To create the two conditions, you need to select a female name and a male name. The apparent ethnicity of the names should be the same. For example, if the female name is Hispanic, then the male name should be Hispanic.

You then want to devise a way to measure the dependent variable. You might ask how likely subjects would be to vote for such a candidate, the extent to which they agree with the candidate's positions, and so on. Be sure to check the effectiveness of the manipulation. For example, after retrieving the description, you might ask them about various characteristics of the candidate, such as age, gender, and occupation.

After you enter data into a MicroCase data set, you can test the hypothesis that the gender of the candidate affects voters' preferences.

PROJECT 5: DOING CONTENT ANALYSIS

In the study described in Chapter 6 in the textbook, John C. Merrill reported the results of a content analysis of *Time* magazine stories on Presidents Truman, Eisenhower, and Kennedy. This project is based on that study.

The purpose of this study is to use content analysis to see if different magazines put their own "spin" on the behavior of the president. To obtain the multiple coders needed to check reliability in content coding, you might work with one or two classmates on this project, if your instructor approves. Otherwise, you will need to arrange for at least one coder in addition to yourself.

Select a current event in which the president was involved. Then you can go to the library and select several newspapers to compare. These might be from different parts of the country. Or you could select two newsmagazines or newspapers from different places on the political spectrum. For example, you might select one "mainstream" magazine such as *Time* or *Newsweek*, and a conservative magazine such as *The National Review* or a liberal magazine such as *The Nation*. Obtain from each publication stories on the selected event. Provide each coder with a copy of the stories, and ask him or her to identify each instance of bias and code it as positive or negative in the following categories:

1. *Attribution bias:* use of a "loaded" verb, such as "barked," "smiled," or "waffled," in place of a neutral verb, such as "said."
2. *Adjective bias:* use of a favorable or unfavorable adjective, such as "disorganized," "boring," "forceful," or "effective." These are subjective, or judgmental, adjectives, in contrast to objective, or neutral, adjectives, such as "blue" sky.
3. *Adverbial bias:* use of a favorable or unfavorable adverb, such as "warmly," "curtly," or "slyly." Frequently, these will be combined with attribution bias, such as in "chatted amiably" or "barked sarcastically."
4. *Outright opinion:* reaching a subjective conclusion rather than stating facts: "His inability to stay focused on the issue has confused his supporters as well as his opponents" or "His forceful presentation put his opponents on the defensive."
5. *Photographic bias:* portraying the president in a positive or negative manner in photographs or cartoons.

In tallying the codes, use only instances of bias on which all coders agree. If one coder cites an instance of negative bias that the other coders ignore, drop that instance. You can then compare the two magazines on each type of bias.

Appendix A: Variable Names and Sources

Note: The full version of the MicroCase program can access the data files provided with this book. However, if you use the full version of MicroCase to move variables from these files into other MicroCase files, or vice versa, you may need to reorder the cases. Also, note that files which have been modified in the full version of MicroCase will not function properly in Student MicroCase.

◆ **SHORT LABEL: NES** ◆

1) AGE
2) AGE CATEGR
3) EDUCATION
4) EDUC CATEG
5) SEX
6) RACE
7) REGION
8) HISPANIC
9) FAMINCOME
10) %50%50%
11) SCHL PRAYR
12) RELIGPOLS
13) RELIGPOLT
14) RELIG IMPOR
15) PRAYFREQ
16) READ BIBLE
17) BIBLEVIEW
18) REL FREQ
19) RELPREF
20) MEM CHURCH
21) CHRST TYPE

22) BORN AGAIN
23) CLINT THRM
24) GORE THERM
25) BUSH THERM
26) MCCAIN THM
27) BRADLEY TH
28) HILARY THM
29) DEMS THRM
30) REPS THRM
31) REL RT THM
32) CONGRS THM
33) ABORTION
34) ABORTN DEM
35) ABORTN REP
36) LATE TERM
37) INTEREST?
38) INTERNET
39) CLINTON
40) LSTN FREQ
41) VOTE96
42) VOTE98?

43) PRESVOTE96
44) CARE HOUSE
45) KNOW HOUSE
46) CONGRESS
47) WHO HSE PT
48) WHO SEN PT
49) WHO GUV PT
50) APPRV REP
51) TERM LIMIT
52) PARTY3
53) PARTY7
54) FOLLOW GOV
55) TALK POLTC
56) PARTY DIFF
57) LIBCON7
58) LIBCON3
59) DEMLIBCON
60) REPLIBCON
61) FINANCES
62) FIN FUTUR
63) ECONOMY

◆ SHORT LABEL: NES cont'd ◆

64) ECON FUTUR
65) STOCKS?
66) PTY CONTRL
67) PARTIES
68) EQUAL RGTS
69) BLEND IN?
70) WOMEN EQL
71) WOMEN DEM
72) WOMEN REP
73) OFF LANG?
74) JOB GUAR
75) SPENDING
76) SPEND DEM
77) SPEND REP
78) AFFIRM ACT
79) POLITKNOW
80) POL INFO

81) FAIR MEDIA
82) SCHL VOUCR
83) STRONG US?
84) STAY HOME
85) IMMIGRATE
86) IMPORTS
87) VIETNAM
88) ENV REGS
89) ENV DEM
90) ENV REPUB
91) DEATH PENT
92) FAMILY VAL
93) TOLER DIF
94) INFIDELITY
95) MAKE DIFFER
96) GOVT ATTN
97) ELECT ATTN

98) COMPLICTD
99) DON'T CARE
100) NO SAY
101) CROOKED?
102) GOVT WASTE
103) G DO RIGHT
104) BIG INTERS
105) NO PARTIES
106) TAKE ADVAN
107) TRUST?
108) RESIGN?
109) IMPEACH?
110) APPR IMPCH
111) IMPR INVST
112) COLLEGE?
113) SURPLUS!
114) POLPARTY!

◆ SHORT LABEL: GSS ◆

1) READ PAPER
2) WATCH TV
3) EVER UNEMP
4) FEAR WALK
5) COMPREND
6) ATTITUDE?
7) HAPPY?
8) SPANK?
9) SOCBAR
10) EVER STRAY
11) SEX FREQ
12) HEALTH
13) MOVERS
14) COP:HIT?
15) COP:ESCAPE
16) COP:ATTACK
17) OWN GUN?
18) YOUR GUN?
19) COURTS?

20) HUNT?
21) WORK IMPOR
22) POLPARTY
23) POLVIEW
24) VOTE IN 92
25) WHO IN 92?
26) VOTE IN 96
27) WHO IN 96?
28) ATHSPK
29) RACSPK
30) COMSPK
31) MILSPK
32) GAYSPK
33) FREESPEAK
34) WELFARE $
35) WELFARE $2
36) HEALTH $
37) HEALTH $2
38) CRIME $

39) CRIME $2
40) ENVIRON $
41) ENVIRON $2
42) BIG CITY $
43) BIG CITY$2
44) DRUGS $
45) DRUGS $2
46) BLACK $
47) BLACK $2
48) SOC.SEC.$
49) ABORT DEF
50) ABORT WANT
51) ABORT HLTH
52) ABORT POOR
53) ABORT RAPE
54) ABORT SING
55) ABORT INDX
56) INTERMAR?
57) WOMEN HOME

◆ SHORT LABEL: GSS cont'd ◆

58) MEN BETTER
59) WOMAN PRES
60) HOUSEWIFE
61) HELP HUSB
62) MARR ROLES
63) SCHOOLPRAY
64) FED GOVT?
65) SUP COURT?
66) CONGRESS?
67) MILITARY?
68) RELIGION?
69) EDUCATION?
70) GOV MED
71) MUCH GOVT
72) GRASS?
73) EXECUTE?
74) RELIGION
75) RELIG ID
76) DENOM
77) R.FUND/LIB
78) BIBLE1
79) BELIEF GOD
80) READ BIBLE

81) REBORN
82) MEDITATE
83) RELPREF S
84) RELPERSON
85) PRAYFREQ
86) AFTERLIFE?
87) MEDIA REL
88) ATTEND
89) SEX
90) REGION4
91) RACE
92) WH/AFRI.AM
93) WORKING?
94) OWN HOME?
95) ZODIAC
96) MARITAL
97) PHONE
98) #SIBS
99) #CHILDREN
100) DEGREE
101) AGE
102) OVER 50?
103) ED YEARS

104) INCOME
105) R.INCOME
106) %50%50%
107) OCCPRST SC
108) DAD OCPRS
109) MOM OCPRS
110) PLACE SIZE
111) URBAN?
112) DEGREE!
113) AGE!
114) EDUCATION!
115) INCOME!
116) R.INCOME!
117) #SIBS!
118) DAD EDUC!
119) MOM EDUC!
120) POLPARTY!
121) POLVIEW!
122) SEX FREQ!
123) RELPREF S!
124) PRAYFREQ!
125) HEALTH!
126) ATTEND!

◆ SHORT LABEL: USA ◆

1) STATE NAME
2) POP 98
3) %WHITE
4) %BLACK
5) %HISPANIC
6) %URBAN
7) DENSITY
8) SYPHILIS
9) INFANT MOR
10) CAR-DEATH
11) SUICIDE
12) MARRIAGE
13) DIVORCE

14) HEALTH INS
15) AIDS
16) CIGSMOKERS
17) %HIGH SCH
18) %COLLEGE
19) POLICE
20) KID ABUSES
21) #SOC.SEC
22) SS BENEFIT
23) %SOCSEC
24) AVG UNEMP$
25) UNEMP$/CAP
26) FOODST#

27) FS$/PER
28) FS$/CAP
29) AVG PAY
30) MED FAM$
31) PER CAP$
32) POV LINE
33) STATE TAX
34) EXPENDITUR
35) EXP$/CAP
36) VOTED98
37) TANF FAM
38) TEEN MOMS
39) IMMIGRANTS

◆ SHORT LABEL: USA cont'd ◆

40) HATECRIMES	74) %REGIST 96	108) BYTE
41) EDUC$/CAP	75) MILES/DRV	109) ROLLSTONE
42) HEALTH/CAP	76) MILES/VHCL	110) R DIGEST
43) PRISON/CAP	77) SOUTHNESS	111) NAT REVIEW
44) ABORTIONS#	78) WESTNESS	112) MOTHR JONE
45) ABORTRATE1	79) WARM WINTR	113) COSMO
46) ABORTRATE2	80) ELEVATION	114) PLAYBOY
47) ABORTRATE3	81) %<5	115) F&STREAM
48) BIRTHRATE	82) %<20	116) CIG TAX
49) %BEER	83) %>64	117) ART GRANTS
50) %WINE	84) %5–9	118) MOTOR VTR
51) ETHGAL/CAP	85) %10–14	119) MED AGE
52) ALCBEV/CAP	86) %15–19	120) REGION
53) $PER PUPIL	87) %2–24	121) VIOLENT CR
54) PRISONERS#	88) %25–29	122) CRIME RATE
55) TOXIC	89) AIDS DEATH	123) %NORELIG
56) LOC$/CAP	90) SHRINKS	124) % JEWISH
57) LOCEX$/CAP	91) %FEM MD	125) % CATHOLIC
58) LOCDBT/CAP	92) VEHICLES	126) % BAPTIST
59) FEDFUNDS	93) CYCLES	127) CH.MEMB
60) VETERANS	94) PICKUPS	128) FOODSTAMPS
61) FDAID-CHFM	95) UNEMPLOY	129) HUNTING
62) FDAID-HOUS	96) FEM UNEMPL	130) FISHING
63) DEM CONG	97) MAL UNEMPL	131) WILDWATCH
64) REPUB CONG	98) % UNION	132) HUNTDAYS
65) PCOMP-CONG	99) COMPUTERS	133) WELFAR/CAP
66) PCOMP-LEG	100) PROP CRIME	134) PLAYBOY#
67) %CLINTON92	101) MURDER	135) AREA
68) %BUSH '92	102) RAPE	136) % FAT
69) %PEROT 92	103) ROBBERY	137) PLASTIC
70) %CLINTON96	104) ASSAULT	138) CHIROPRACT
71) %DOLE 96	105) BURGL	139) FEM-LEGIS
72) %PEROT 96	106) LARCENY	140) NDM 95–96
73) %VOTED 96	107) AUTO THEFT	

◆ SOURCES ◆

NES — AMERICAN NATIONAL ELECTION STUDY, 1998

The NES data file is based on selected variables from the 1998 American National Election Study provided by the National Election Studies, Institute for Social Research at the University of Michigan and the Inter-university Consortium for Political and Social Research. The principal investigators were Virginia Sapiro and Steven Rosenstone. Kathy Cirksena was the Project Manager and Michael Horvath was the Study Manager for the National Election Studies.

GSS — NATIONAL OPINION RESEARCH CENTER GENERAL SOCIAL SURVEY, 1998

The GSS data file is based on selected variables from the National Opinion Research Center (University of Chicago) General Social Survey for 1998. James A. Davis is the Principal investigator and Tom W. Smith is the Director and Co-Principal Investigator.

USA — THE 50 STATES OF THE U.S.

The data in the USA file are from a variety of sources, which are indicated in the long labels for the variables. Some of the sources are abbreviated as follows.

CENSUS: The summary volumes of the U.S. Census for the indicated year.

CHURCH: *Churches and Church Membership in the United States*, published every 10 years by the Glenmary Research Center, Atlanta, for the indicated year.

FEC: Federal Election Commission for the indicated year.

KOSMIN: Kosmin, Barry A. 1991. *Research Report: The National Survey of Religious Identification*, New York: CUNY Graduate Center.

S.A.: *Statistical Abstract of the United States* for the year indicated.

SR: *State Rankings* (Morgan Quitno Corp., Lawrence, KS) for the indicated year.

UCR: *The Uniform Crime Reports* for the indicated year.

WA: *The World Almanac* for the indicated year.

The USA file also uses data from *USA Today*, 6/23/97.

Variables with no source shown are from the U.S. Census publications.

Appendix B: Student Questionnaire

Please answer the following questions by placing a mark in the appropriate blank or by writing in the requested information.

1. I really like science and math courses.
   - 1___Strongly agree
   - 2___Agree
   - 3___Disagree
   - 4___Strongly disagree

2. I am not sure that college is worth all the bother.
   - 1___Strongly agree
   - 2___Agree
   - 3___Disagree
   - 4___Strongly disagree

3. My career plans after I finish college are:
   - 1___Very definite
   - 2___Fairly clear
   - 3___At the "maybe" stage
   - 4___Still pretty much undecided

4. What is your present year in college?
   - 1___First
   - 2___Second
   - 3___Third
   - 4___Fourth
   - 5___Fifth or more

5. Do you live at home, in a dorm, or where?
   - 1___At home with my parent(s)
   - 2___In my own apartment or house
   - 3___In a dorm
   - 4___In a sorority
   - 5___In a fraternity

6. During an average week, how many hours do you spend studying for college?_____(write in number)

7. What is your GPA (Grade Point Average)?_____(write in number) If this is your first term in college, report your high school GPA.

8. Do you belong to a fraternity or sorority, whether as a pledge or as an active member?
    1___Yes
    2___No

9. If you had to choose between a course in literature and a course in science, which would you probably select?
    1___Literature
    2___Science

10. What is your age?_____

11. Have you ever received a ticket, or been charged by the police, for a traffic violation—other than illegal parking?
    1___Yes
    2___No

12. Were you ever picked up, or charged, by the police for any other reason, whether or not you were guilty?
    1___Yes
    2___No

13. Have you ever shoplifted?
    1___Yes
    2___No

14. Whether or not you ever have drunk alcoholic beverages such as liquor, wine, or beer, do you do so now or are you a total abstainer?
    1___Drink now
    2___Abstain now

15. Have you EVER tried marijuana?
    1___Yes, in the past year
    2___Yes, but not in the past year
    3___Never

16. Have you EVER tried cocaine (crack, rock, freebase)?
    1___Yes, in the past year
    2___Yes, but not in the past year
    3___Never

17. During the past year, have you been nauseated or vomited due to your drinking or drug use?
  1___Yes
  2___No

18. Have you ever cheated on an exam?
  1___Yes, very often
  2___Yes, quite often
  3___Yes, a few times
  4___Yes, but only once
  5___No, not ever

19. IF you have ever cheated on an exam, was that in high school or in college?
  1___In college
  2___In high school
  3___Both

20. Do you agree that people ought to have the right to end their own lives anytime they arc tired of living?
  1___Strongly agree
  2___Agree
  3___Disagree
  4___Strongly disagree

21. Do you favor or oppose the death penalty for persons convicted of murder?
  1___Favor
  2___Oppose

Do you approve or disapprove of abortions under each of the following circumstances:

22. If the woman is not married and does not want to marry the man.
  1___Approve
  2___Disapprove

23. If the woman's own health is seriously endangered by the pregnancy.
  1___Approve
  2___Disapprove

24. If the woman is married but doesn't want any more children.
  1___Approve
  2___Disapprove

25. If the woman simply wants an abortion for any reason at all.
  1___Approve
  2___Disapprove

26. Do you smoke?

1___No

2___Yes

27. Do you have a computer?

1___Yes

2___No

Do you think the government ought to spend more money, less money, or the current level on each of the following:

28. On welfare?

1___Spend more

2___Spend less

3___Spend at the current level

29. On defense?

1___Spend more

2___Spend less

3___Spend at the current level

30. On space exploration?

1___Spend more

2___Spend less

3___Spend at the current level

31. On education?

1___Spend more

2___Spend less

3___Spend at the current level

32. The press often reports predictions about the future by people such as Jean Dixon, who claim to have psychic powers. Do you think some people do have such powers?

1___I am certain some people do have psychic powers.

2___I think some people probably have psychic powers.

3___I tend to doubt that anyone is a psychic.

4___I am certain this is all nonsense.

33. How much confidence do you place in astrology—the theory that the position of the stars and planets in relation to our birthdays has a lot to do with what we are like and what will happen to us?

1___A lot of confidence
2___Some confidence
3___Not much confidence
4___No confidence

34. Whom did you favor in the 1996 presidential election?

1___Dole
2___Clinton
3___Perot
4___Other

35. About how often do you attend religious services?

1___More than once a week
2___About once a week
3___At least once a month
4___At least twice a year
5___Seldom
6___Never

36. What is your religious preference?

1___Catholic
2___Protestant
3___Jewish
4___Other
5___None

37. Would you say that you are a religious person or that you are not?

1___Very religious
2___Somewhat religious
3___Not very religious
4___Not religious

38. When you were in high school, did you participate in an organized sport that involved competition with other schools?

1___Yes
2___No

39. Are you employed?

1___Yes, full-time
2___Yes, part-time
3___No

40. What is your current marital status?
1___Single (never married)
2___Married
3___Divorced or separated
4___Widowed

41. When you were 16, with whom were you living?
1___Both parents
2___One parent and a stepparent
3___My mother
4___My father
5___Others

42. Thinking about your parents, or the people with whom you lived during high school, compared with other American families, would you say their income was below average or above?
1___Far below average
2___Below average
3___Average
4___Above average
5___Far above average

43. Thinking about the home you lived in when you were in high school, about how many hardcover books were in the house?
1___0–10
2___11–25
3___26–75
4___76–100
5___101–200
6___201–500
7___More than 500

44. What do you think is the ideal number of children for a family to have?
0___None
1___One
2___Two
3___Three
4___Four
5___Five
6___Six or more

45. Race/ethnicity:
    1___White (Anglo)
    2___African American
    3___Asian American
    4___Hispanic American
    5___Native American
    6___Pacific Islander
    7___Other_____(write in)

46. Sex:
    1___Female
    2___Male

47. Were you born in the United States?
    1___Yes, in this state
    2___Yes, in another state
    3___No

48. Do you agree or disagree that a preschool child is likely to suffer if his or her mother works?
    1___Agree
    2___Disagree

49. It might be better for everyone if the husband takes care of earning a living and the wife takes care of the home and the children.
    1___Agree
    2___Disagree

50. Do you usually wear one or more rings on your fingers?
    1___Usually wear more than one ring
    2___Usually wear one ring
    3___Sometimes wear a ring
    4___Seldom wear a ring
    5___Never wear a ring

51. At present, do you have your own car?
    1___Yes
    2___No

52. If you had to be one or the other, would you rather be a dog or a cat?
    1___Dog
    2___Cat

53. Would you rate yourself as overweight, about right, or underweight?
   1___Quite overweight
   2___Somewhat overweight
   3___About right
   4___Somewhat underweight
   5___Quite underweight

54. How likely do you think it is that during your lifetime you will suffer a significant hair loss?
   1___Very likely
   2___Somewhat likely
   3___Not very likely
   4___Very unlikely

55. Have you ever used a sewing machine?
   1___Often
   2___Sometimes
   3___Once or twice
   4___Never

Please write in an answer to each of the following questions, estimating the answer to the best of your knowledge

56._____Estimate the world's total population (in billions).

57._____Estimate the percentage of the U.S. population who are African American.

58._____Estimate the percentage of Americans who are the victims of violent crime in any given year.

59._____Estimate the American family's median income in dollars per year.

60._____Estimate the percentage of Americans of voting age who voted in the 2000 presidential election.

61._____Estimate the percentage of the total popular vote received by George W. Bush in the 2000 presidential election.

62._____Estimate the percentage of American adults who drink alcoholic beverages.

LICENSE AGREEMENT FOR WADSWORTH GROUP/THOMSON LEARNING

CD-ROM MATERIALS

You the customer, and Wadsworth Group/Thomson Learning incur certain benefits, rights, and obligations to each other when you open this package and use the materials it contains. BE SURE TO READ THIS LICENSE AGREEMENT CAREFULLY, SINCE BY USING THE CD-ROM YOU INDICATE YOU HAVE READ, UNDERSTOOD, AND ACCEPTED THE TERMS OF THIS AGREEMENT.

Your rights:

1. You enjoy a nonexclusive license to use the enclosed materials on a single computer that is not part of a network or multimachine system in consideration of the payment of the required license fee, (which may be included in the purchase price of an accompanying print component), and your acceptance of the terms and conditions of this agreement.
2. You own the CD-ROM disc on which the program/data is recorded, but you acknowledge that you do not own the program/data recorded on the CD-ROM. You also acknowledge that the program/data is furnished "AS IS," and contains copyrighted and/or proprietary and confidential information of Wadsworth Group/Thomson Learning and its licensors.
3. If you do not accept the terms of this license agreement you must not install the CD-ROM and you must return the CD-ROM within 30 days of receipt with proof of payment to Wadsworth Group/Thomson Learning for full credit or refund.

These are limitations on your rights:

1. You may not copy or print the program/data for any reason whatsoever, except to install it to a hard drive on a single computer, unless copying or printing is expressly permitted in writing or statements recorded on the CD-ROM.
2. You may not revise, translate, convert, disassemble, or otherwise reverse engineer the program/data.
3. You may not sell, license, rent, loan, or otherwise distribute or network the program/data.
4. You may not export or reexport the CD-ROM, or any component thereof, without the appropriate U.S. or foreign government licenses.

Should you fail to abide by the terms of this license or otherwise violate Wadsworth Group/Thomson Learning's rights, your license to use it will become invalid. You agree to destroy the CD-ROM immediately after receiving notice of Wadsworth Group/Thomson Learning's termination of this agreement for violation of its provisions.

U.S. Government Restricted Rights

The enclosed multimedia, software, and associated documentation are provided with RESTRICTED RIGHTS. Use, duplication, or disclosure by the Government is subject to restrictions as set forth in subdivision (c)(1)(ii) of the Rights in Technical Data and Computer Software clause at DFARS 252.277.7013 for DoD contracts, paragraphs (c) (1) and (2) of the Commercial Computer Software-Restricted Rights clause in the FAR (48 CFR 52.227-19) for civilian agencies, or in other comparable agency clauses. The proprietor of the enclosed multimedia, software, and associated documentation is Wadsworth Group/Thomson Learning, 10 Davis Drive, Belmont, California, 94002.

Limited Warranty

Wadsworth Group/Thomson Learning also warrants that the optical media on which the Product is distributed is free from defects in materials and workmanship under normal use. Wadsworth Group/Thomson Learning will replace defective media at no charge, provided you return the Product to Wadsworth Group/Thomson Learning within 90 days of delivery to you as evidenced by a copy of your invoice. If failure of disc(s) has resulted from accident, abuse, or misapplication Wadsworth Group/Thomson Learning shall have no responsibility to replace the disc(s). THESE ARE YOUR SOLE REMEDIES FOR ANY BREACH OF WARRANTY.

EXCEPT AS SPECIFICALLY PROVIDED ABOVE, WADSWORTH GROUP/THOMSON LEARNING AND THE THIRD PARTY SUPPLIERS MAKE NO WARRANTY OR REPRESENTATION, EITHER EXPRESSED OR IMPLIED, WITH RESPECT TO THE PRODUCT, INCLUDING ITS QUALITY, PERFORMANCE, MERCHANTABILITY, OR FITNESS FOR A PARTICULAR PURPOSE. The Product is not a substitute for human judgement. Because the software is inherently complex and may not be completely free of errors, you are advised to validate your work. IN NO EVENT WILL WADSWORTH GROUP/THOMSON LEARNING OR ANY THIRD PARTY SUPPLIERS BE LIABLE FOR DIRECT, INDIRECT, SPECIAL, INCIDENTAL, OR CONSEQUENTIAL DAMAGES ARISING OUT OF THE USE OR INABILITY TO USE THE PRODUCT OR DOCUMENTATION, even if advised of the possibility of such damages. Specifically, Wadsworth Group/Thomson Learning is not responsible for any costs including, but not limited to, those incurred as a result of lost profits or revenue, loss of use of the computer program, loss of data, the costs of recovering such programs or data, the cost of any substitute program, claims by third parties, or for similar costs. In no case shall Wadsworth Group/Thomson Learning's liability exceed the amount of the license fee paid. THE WARRANTY AND REMEDIES SET FORTH ABOVE ARE EXCLUSIVE AND IN LIEU OF ALL OTHERS, ORAL OR WRITTEN, EXPRESS OR IMPLIED. Some states do not allow the exclusion or limitation of implied warranties or limitation of liability for incidental and consequential damage, so that the above limitations or exclusion many not apply to you.

This license is the entire agreement between you and Wadsworth Group/Thomson Learning and it shall be interpreted and enforced under California law. Should you have any questions concerning this License Agreement, write to Technology Department, Wadsworth Group/Thomson Learning, 10 Davis Drive, Belmont, California, 94002.